Storm Surge Forecasting and Future Projection in Practice

This accessible shortform book describes storm surge forecasting to enable port managers and practitioners to forecast these and mitigate their effects. This is particularly useful as global warming increases the severity of typhoons, particularly windstorms and storm surge disasters, globally.

The authors first summarize the current status of typhoons and storm surges in practice. They also present a combination of the latest findings at the research level and at the practical level. Throughout the book, the authors carefully explain the use and limitations of empirical typhoon models that practitioners should learn from, including statistical, numerical, probabilistic, data-driven and coastal vulnerability models. They also explore artificial neural networks and convolutional neural networks and their use in such models. Finally, the book describes the potential for further development of empirical typhoon models (such as future climate experiments).

This book is a vital resource that enables port managers to make effective and informed decisions when conducting storm surge forecasting in practice. It also contains useful insights for civil engineering students, especially those studying coastal engineering.

Masaya Toyoda is an Assistant Professor at the Coastal Engineering Laboratory, Graduate School of Engineering, Toyohashi University of Technology, Aichi, Japan. He attained his Bachelor's, Master's and Doctoral degrees from Gifu University in 2015, 2017 and 2020, respectively.

Nobuki Fukui is an Assistant Professor at the Disaster Prevention Planning Laboratory, Graduate School of Engineering, Tottori University, Tottori, Japan. He completed his Bachelor's, Master's and Doctoral degrees from Kyoto University in 2017, 2019 and 2022, respectively.

Storm Surge Forecasting and Future Projection in Practice

Scope of Application of Empirical Typhoon Models

Masaya Toyoda and Nobuki Fukui

CRC Press
Taylor & Francis Group
Boca Raton London New York

CRC Press is an imprint of the
Taylor & Francis Group, an **informa** business

First edition published 2025
by CRC Press
2385 NW Executive Center Drive, Suite 320, Boca Raton FL 33431

and by CRC Press
4 Park Square, Milton Park, Abingdon, Oxon, OX14 4RN

CRC Press is an imprint of Taylor & Francis Group, LLC

© 2025 Masaya Toyoda and Nobuki Fukui

Library of Congress Cataloging-in-Publication Data
Names: Toyoda, Masaya, author. | Fukui, Nobuki, author.
Title: Storm surge forecasting and future projection in practice : scope of
application of empirical typhoon models / Masaya Toyoda and Nobuki Fukui.
Description: First edition. | Boca Raton FL : CRC Press, 2025. | Includes
bibliographical references and index.
Identifiers: LCCN 2024020469 | ISBN 9781032765099 (hardback) |
ISBN 9781032765105 (paperback) | ISBN 9781003478782 (ebook)
Subjects: LCSH: Storm surges—Mathematical models. | Typhoons—Forecasting. |
Neural networks (Computer science)
Classification: LCC GC225 .T68 2025 | DDC 551.46/3015118—dc23/eng20240723
LC record available at https://lccn.loc.gov/2024020469

ISBN: 9781032765099 (hbk)
ISBN: 9781032765105 (pbk)
ISBN: 9781003478782 (ebk)

DOI: 10.1201/9781003478782

Typeset in Minion
by codeMantra

Contents

Acknowledgment

This research work was supported by two JSPS Research Fellow Grants (No. 17J04771, No. 20J00218, 23K13411, 24K17357).

CHAPTER 1

Introduction

Recently, disasters caused by extreme weather events have been reported globally owing to the progression of global warming. According to theWorld Meteorological Organization (WMO)'s State of the Global Climate 2022, the year 2022 is expected to see damage from heat waves, wildfires, cold waves, floods, tropical cyclones (TCs) and severe storms. WMO State of the Global Climate 2022 summarizes the damage caused by heat waves, wildfires, cold waves, floods, TCs and severe storms. The TC activity was near normal or below normal in most areas except in the South Indian Ocean. On the other hand, the scale of each disaster was large, with two major hurricanes making a landfall in the North Atlantic in September. For example, Fiona crossed the Dominican Republic, Puerto Rico and the Turks and Caicos Islands, causing extensive flooding and widespread power outages. This extratropical storm made a landfall in Nova Scotia, Canada, at 932.7 hPa, the lowest sea-level pressure ever recorded. The coastal areas, especially in southwestern Newfoundland, experienced a significant storm surge and wave inundation. Hurricane Ian made a landfall in southwestern Florida with Category 4 force, causing major storm surge inundation of coastal areas. It was the fourth strongest hurricane on record to make a landfall in Florida. The WMO article stated that TCs were below average in activity in the Northwest Pacific, but the two tropical storms that hit the Philippines had a destructive impact. As described above, even in 2022, when TC's activity was lower than normal year, TCs caused damage in many areas.

The first example of a TC disaster that caused the most serious damage in the world in the 21st century is Hurricane Katrina, which hit

DOI: 10.1201/9781003478782-1

southeastern US at the end of August 2005. The storm surge caused by this hurricane submerged approximately 80% of New Orleans, causing the worst damage in US history (National Institute for Land and Infrastructure Management, 2005; National Research Institute for Earth Science and Disaster Resilience 2006). Cyclone Nargis, which hit central and southern Myanmar in May 2008, is said to be the worst natural disaster in the history of Myanmar, with 150,000 people reported dead or missing (Cabinet Office, 2008). Typhoon Haiyan, which hit the central Philippines in November 2013, had a minimum central pressure of 895 hPa and maximum wind speed of 65 m/s. It caused a storm surge with a maximum tide level of over 6 m in San Pedro Bay, Leyte Island, causing extensive damage (Lander et al., 2014). In Japan, the approach and passage of Typhoon No. 21 in 2018 brought extremely strong and fierce winds, intense rain from western to northern Japan and a significant storm surge (3.29 m) that exceeded the previous maximum tide level in Osaka Bay (JMA, 2018a). The typhoon also caused 14 deaths and 980 injuries (including serious and minor) (as of 1 April 2019; Fire and Disaster Management Agency, 2019). In response to the abovementioned damage from strong winds and storm surge, the General Insurance Association of Japan (GIAJ) announced that the total amount of claims paid for Typhoon No. 21 exceeded 1,067 billion yen (US$9.7 billion) (announced on May 20, 2019). Other storm surge disasters were caused by Typhoon No. 16 in 2004, Typhoon No. 18 in 2009 and Typhoon No. 16 in 2012. Typhoons No. 9, 10 and 11 in 2016 caused extensive damage to crops in the Tohoku and Hokkaido regions. The Japan Meteorological Agency (JMA) reported 89 cases in the 24-year period from 2000 to 2023 (JMA, 2023). Among them, 47 cases were directly or indirectly caused by TCs (typhoons), indicating that more than half of the meteorological disasters that will occur in Japan in the 21st century and beyond will be related to typhoons.

As described above, large-scale disasters caused by TCs have occurred in the Pacific and Atlantic regions over the past 20 years since 2000. Additionally, not only in developing countries, but also in countries with relatively advanced disaster countermeasures against TCs, such as Japan and the US, damage has occurred, and there is an urgent need to improve disaster prevention and mitigation systems against future extreme weather disasters. Moreover, many recent weather disasters described above were caused by "natural phenomena that exceeded expectations" or "external forces that exceeded design standards," resulting in extensive damage, which can be attributed to changes in the natural and social environments due to global warming.

1.1 WHAT IS A TROPICAL CYCLONE?

Internationally, all TCs are classified into three broad categories according to their intensity based on the maximum wind speed in the area: tropical depression, tropical storm and developed TC. TCs have different names in different regions. They are known as hurricanes in the Atlantic Ocean and North Pacific Ocean east of the date line, typhoons in the Northwest Pacific Ocean and cyclones in the Indian Ocean and southern Pacific Ocean. The Saffir–Simpson hurricane wind scale is a measure of hurricane intensity. This index classifies hurricane intensity into five categories. Category 1 is for hurricanes with maximum wind speeds of 33–42 m/s, Category 2 for 43–49 m/s, Category 3 for 50–58 m/s, Category 4 for 58–70 m/s and Category 5 for hurricanes with wind speeds of >70 m/s. In recent years, the Joint Typhoon Warning Center designated typhoons with maximum one-minute mean wind speeds exceeding 67 m/s as super typhoons. In other regions, this index is not directly used owing to the different averaging method of maximum wind speeds.

TCs have the characteristics of moving under the influence of the winds above and the pressure pattern of the surrounding area. For example, a typhoon in Japan tends to move westward at low latitudes and then moves northward around the Pacific High, and when it reaches mid to high latitudes, it tends to move northeastward at a high speed owing to the westerly winds overhead. The intensity of typhoons is generally estimated using the Dvorak method, which is based on the correspondence between the cloud patterns observed by meteorological satellites and the actual central pressure and maximum wind speed. TCs develop using the heat released when water vapor supplied from the warm ocean surface condenses to form cloud particles. However, as it moves, it continuously loses energy because of friction with the sea surface and ground, and if the energy supply was to cease, it would dissipate in two to three days. Additionally, as it approaches near Japan, cold air begins to flow into the sky, and it gradually loses its original typhoon characteristics and turns into an extratropical cyclone (EC). Alternatively, it may become a TC as the supply of heat decreases, and therefore decays. The reason a TC that has made a landfall rapidly decays is because the supply of water vapor is discontinued and energy is lost because of the friction between land and water.

TCs bring stormy winds and heavy rainfall at the time of approach and landfall. This often causes river flooding, coastal storm surges and tidal waves in the areas where the cyclone strikes. Floods are caused by heavy rainfall, while storm surges and high waves are caused by winds. When a strong TC hits, storm surges and high waves occur simultaneously, which can be extremely critical in coastal areas. It has been indicated that the intensity of TCs has been increasing recently, and there is concern that TC disasters may become more severe. In this book, the storm surge associated with TCs is especially targeted, and the basic knowledge of storm surge is described.

1.2 STORM SURGE ASSOCIATED WITH TCS

The sea surface periodically cycles between high and low tides once or twice a day owing to the gravitational pull of the moon and sun, and the height of the sea surface (tide level) is called the astronomical tide level. Astronomical tide levels can be calculated in advance, and it is now possible to calculate astronomical tide levels globally via the Internet. However, when typhoon winds blow from offshore toward the coast, seawater is blown toward the coast, causing a rise in sea level near the coast. This is called the blow-in effect. This effect is expressed as the balance between wind stress and sea-surface gradient force as:

$$g\frac{\partial \eta}{\partial x} = \frac{\tau}{\rho h} \tag{1.1}$$

$$\eta = \int_0^L \frac{\tau}{\rho g h} dx = \frac{\tau}{\rho g h} \cdot L \tag{1.2}$$

where τ is the wind stress, L is the horizontal distance from the point of interest to the shore, h is the water depth, ρ is the seawater density and g is the acceleration due to gravity. In the case of a V-shaped bay, the sea level rises in the deeper part of the bay, because the narrower the bay, the higher the increase in sea level. Additionally, the sea level rises when the atmospheric pressure becomes low because of an approaching typhoon. This is called the "suction effect." In the open ocean, a 1 hPa drop in atmospheric pressure increases the sea level by about 1 cm. The suction effect is expressed based on the static balance between the sea level and atmospheric pressure as:

$$\rho g \Delta h \cdot S = \Delta p \cdot S \tag{1.3}$$

where Δp is the amount of pressure drop [hPa], which is generally calculated as "1013 hPa—the central pressure of the TC." S is the area to be calculated, and, as it is on both sides, it can be divided to form Δh:

$$\Delta h = \frac{\Delta p}{\rho g}. \tag{1.4}$$

Assuming $\Delta p = 1.0$ [hPa], we obtain:

$$\Delta h = \frac{1.0 \left[\text{hPa} \right]}{\left(1.0 \left[\text{g/cm}^3 \right] \times 9.8 \left[\text{m/s}^2 \right] \right)} = 1.0 [\text{cm}] \tag{1.5}$$

The sea level is calculated as follows. For example, if a typhoon with a central pressure of 950 hPa arrives at a point where the pressure was previously 1,000 hPa, the sea level near the center of the typhoon rises by approximately 50 cm, and the sea level around the center rises in accordance with the pressure. The rise in sea level caused by the blowing-in and sucking-up effects is called a storm surge (actual tide level = astronomical tide level + storm surge deviation). In most cases, the effect of the blow-in effect is greater.

1.3 RECENT EXAMPLES OF DISASTERS CAUSED BY TCS AND STORM SURGES

In this section, the storm surges associated with TCs described in the previous sections are described in terms of actual cases that have occurred in recent years. Three cases since 2010 are described here.

1.3.1 Typhoon Haiyan (2013)

Haiyan is a typhoon that formed at 0:00 UTC on 4 November 2013 in the equatorial region of 6.1° N, 152.2° E. It developed rapidly from the onset and was estimated to have developed to a maximum wind speed of 65 m/s at 12:00 UTC on 7 November with a minimum central pressure of 895 hPa (JMA, 2013; Lander et al., 2014). As it crossed the central Philippines while maintaining its ferocious strength, it caused extensive damage to the island of Leyte, especially in the area around San Pedro Bay. In Tacloban, the city located at the end of San Pedro Bay, the most severely damaged area, a damage survey revealed that a massive storm surge with a maximum tide level of over 6.0 m occurred (Tajima et al., 2014). Unfortunately, sufficient meteorological observation data was not obtained owing to the breakdown of local observation equipment as the typhoon approached. According to the JMA Best Track, Haiyan maintained a central pressure below 900 hPa for a relatively long period from the 7th to the 8th. The reason for this is thought to be the high sea-surface temperatures and La Niña conditions in the ocean around the Philippines at that time. This resulted in higher-than-normal ocean heat storage and favorable conditions for the growth of typhoons. For example, Lin et al. (2014) suggested that water temperatures of 26°C or higher were maintained deep below the sea surface at the time and that Haiyan was moving at a faster speed, which contributed to its stronger typhoon intensity. Toyoda et al. (2015) conducted sensitivity experiments for Haiyan by varying the initial value of the ocean mixing layer thickness using a 3-km mesh high-resolution typhoon model. The results showed that when the initial value of the ocean mixing layer thickness was set to 50 m, the pressure was 897.1 hPa, which was equivalent to the lowest observed central pressure (895 hPa). This value is approximately 20 m higher than the climatological value. Therefore, it can be concluded that sea-surface temperature and ocean heat storage were favorable conditions for the typhoon. The typhoon made a landfall in the Philippines in the early morning of 8 November and moved north-westward and westward, finally moving out to sea in the evening (Japan Meteorological Agency, 2013).

According to the damage estimates for Haiyan by the Philippine government, 6,190 people were killed, 1,785 people were missing and the total number of affected people was estimated to be over 14.1 million (as of January 2014). Furthermore, in a town on the west coast of Leyte Gulf that was severely damaged by the storm surge, 90% of the buildings were reported to be completely destroyed (Japan Broadcasting Corporation, 13 November 2013 article). A reason for such a major disaster in the Philippines, a typhoon-prone country, is the lack of understanding of a "storm surge" among the people. It is believed that many local people did not understand the warning of "surge," meaning storm surge, and were unable to respond to the disaster.

1.3.2 Typhoon Jebi (2018)

Jebi, which originated near Minamitorishima Island on 28 August 2018, moved northwestward south of Japan, turned northward on 3 September and made a landfall in the southern Tokushima Prefecture at around 12:00 pm on 4 September with very strong force. For the first time in 25 years, a typhoon with a "very strong" force (maximum wind speed of 45–54 m/s) made a landfall in Japan. The typhoon made a landfall again at Kobe City, Hyogo Prefecture, at around 14:00 on the 4th, crossed the Kinki region, moved northward across the Sea of Japan, and turned into an EC at 9:00 on the 5th in the Mamiya Strait. As the typhoon approached and passed over western and northern Japan, it brought extremely strong winds, fierce winds and heavy rainfall to the region, and a significant storm surge (3.29 m), exceeding the previous maximum tide level, occurred in Osaka Bay (Japan Meteorological Agency, 2018b). The typhoon caused 14 deaths and 954 injuries (including serious and minor) (as of 1 April 2019; Fire and Disaster Management Agency, 2019). In response to the damage caused by the strong winds and storm surge described above, GIAJ announced that the total amount of claims paid because of Typhoon No. 21 exceeded 1.067 trillion yen (announced on 20 May 2019). The Osaka Prefecture accounted for 607.7 billion yen, or about 60% of the total amount paid prefecture wise (GIAJ, 2019).

The typhoon caused maximum wind speeds of 48.2 m/s and 55.3 m/s at Cape Muroto in Muroto City, Kochi Prefecture, and 46.5 m/s and 58.1 m/s wind speeds at Kansai International Airport. Severe winds were also observed in the Shikoku and Kinki regions. The typhoon caused the highest maximum wind speeds ever recorded at 53 locations nationwide, and the maximum instantaneous wind speeds at 100 locations.

The storm surge recorded a maximum tide height of 3.29 m in Osaka City, Osaka Prefecture, and 2.33 m in Kobe City, Hyogo Prefecture, exceeding the previous maximum tide height recorded by the second Muroto typhoon. These storms, tidal waves and high waves caused flooding of the runways of Kansai International Airport, resulted in canceled flights and ships, suspended train service and damaged lifelines, such as water, power and telephone.

Jebi caused a storm surge that surpassed the previous maximum tide levels at several locations, including Osaka and Kobe. Osaka was hit by a storm surge of more than 3 m, which combined with the tidal waves to flood the runway of Kansai International Airport and caused extensive storm surge damage. Later, the Kinki Regional Development Bureau of the Ministry of Land, Infrastructure, Transportation, and Tourism announced that the minimum pressure and average maximum wind speed were at the same scale level as those of Typhoon Muroto, Typhoon Jane and Typhoon No. 2, which caused extensive damage along the Osaka Bay coast previously, and that the path was almost identical to that of Typhoon No. 2 (Osaka Large-Scale Urban Flood Countermeasures Committee, 2018).

Moreover, the investigation results of the storm surge disaster investigation team conducted by the Japan Society of Civil Engineers (JSCE) Coastal Engineering Committee indicated that the maximum tide level of the storm surge itself was equivalent to the expected maximum tide level, but the addition of high waves may have caused wave overtopping, resulting in inundation. Furthermore, the typhoon traveled at a relatively high speed (approximately 70 km/h at landfall), which tended to cause large tide-level deviations in the bay (JSCE, 2018; Mori et al., 2019). On the other hand, according to the latest Osaka Large-Scale Urban Flooding Countermeasure Guidelines (released in March 2018), the path shifted to the west of this case and the force was assumed to be 900 hPa. Therefore, the magnitudes of the typhoon and storm surge were within the expected levels. However, at the mouth of the Yodo River, flooding caused by rain and storm surge caused by strong winds exceeded the flooding hazard level. In Kobe, where the storm surge was similar to that in Osaka, the waterfront area was inundated. Furthermore, in Osaka, although flood damage in the city was prevented, 7,800 people were stranded at Kansai International Airport for more than 24 h, and various other areas were damaged (The Asahi Shimbun, 2018). In the future, the combination of flooding and storm surge caused by typhoon-induced heavy rainfall and wind storms is expected to cause compound flooding disasters near the mouths of large rivers, such as the Yodo River.

1.3.3 Hurricane Ian (2022)

Ian, which originated on 19 September 2022, increased in intensity as it moved northward across the Caribbean Sea, making a landfall near La Coloma in the Pinar del Rio Province, Cuba, as a Category 3 hurricane. Ian weakened slightly as it passed over Cuba, but on 28 September, it strengthened further, reaching a Category 5 peak for a short period. It then made a landfall at the barrier island of Cayo Costa, Florida, with Category 4 strength. The minimum central pressure was estimated to be 937 hPa and the maximum wind speed was 71.4 m/s (1 min average), with maximum inundation depths of 3–4.6 m at Fort Myers Beach and Estero Island.

Verisk that the world's leading risk analysis firm (https://www.verisk.com/newsroom/verisk-estimates-industry-insured-losses-for-hurricane-ian/), estimated the damage by Hurricane Ian as follows. The insured losses were estimated to be between US$42 billion and US$57 billion. In terms of human casualties, 161 people died, including 5 in Cuba, 150 in Florida, 5 in North Carolina and 1 in Virginia, making Florida's death toll the highest since the hurricane in 1935. The cities of Fort Myers Beach, Cape Coral and Naples were particularly hard hit, with millions of people without power and many residents forced to take shelter on rooftops. Millions of people lost power, forcing many residents to take shelter on rooftops. Sanibel Island, Fort Myers Beach and Pine Island were severely damaged by strong winds and an accompanying storm surge upon landfall, with nearly all standing structures collapsing and the Matlacha Bridge connecting the Sanibel Causeway and Pine Island collapsing. Ian caused the most deaths and economic losses of any hurricane in 2022. In the North Atlantic region, Ian occurred just weeks after Hurricane Fiona, the strongest storm ever to hit Canada, and was the second major disaster in a row to occur in September, the top season for TC activity.

1.4 CLIMATE CHANGE, TCS AND STORM SURGES

The relationship between global warming, TCs and storm surges is discussed in this section. The Intergovernmental Panel on Climate Change (IPCC) AR6 report on global warming states (IPCC, 2021), "There is no doubt that human influence has warmed the atmosphere, oceans, and land since the pre-industrial era," and the likely range of anthropogenic warming was calculated to be 0.8–1.3°C. The report also mentions that "the number of Category 3–5 TCs is likely to have increased over the past 40 years." It suggests that the percentage of strong TCs and peak wind speeds may increase globally as global warming progresses.

The relationship between TCs and global warming has been studied worldwide (Wu and Wang, 2004; Webster et al. 2005; Oouchi et al. 2006; Yoshimura et al. 2006; Murakami and Wang, 2010; Knutson et al., 2010, 2015; Murakami et al., 2012a, 2012b; Christensen et al., 2013; Tsuboki et al., 2015; Kanada et al., 2017). These studies were mainly based on the frequency of occurrence (Yoshimura et al., 2006; Murakami et al., 2012a), intensity (Webster et al., 2005; Knutson et al., 2010, 2015; Murakami et al., 2012b; Christensen et al., 2013), location and path (Wu and Wang, 2004; Murakami and Wang, 2010) and velocity (Yamaguchi et al., 2020). In this section, we review the methods used and the main results obtained from these studies.

First, as a study focusing on the frequency of occurrence, Yoshimura et al. (2006) conducted a future climate experiment using a general circulation model (GCM) with a horizontal resolution of 110 km, three types of sea-surface temperature (SST) rise patterns and two types of cumulus parameterizations assuming warming. The results show that the SSTs are expected to increase. The results show that the frequency of tropical storms (maximum wind speeds of 17 m/s or higher) decreases by 9.0–18.4% globally as SST increases. Murakami et al. (2012a) evaluated changes in TC frequency at the end of the 21st century under the A1B scenario of the IPCC Fourth Assessment Report using a 60-km mesh GCM. They reported that the global frequency of TCs will decrease by 5–35%.

Numerous studies focused on the relationship between global warming and TC intensity (Webster et al., 2005; Oouchi et al., 2006; Murakami et al., 2012b; Christensen et al., 2013; Knutson et al., 2015). Webster et al. (2005) studied the statistics of hurricanes that occurred in the 30 years since 1970. They found that the percentage of hurricanes with intensity of Category 4 (maximum wind speed of 59 m/s per minute on average) or higher was approximately 20% in the 1970s, while it increased to about 35% after the 1990s. The percentage of hurricanes with winds of Category 4 (maximum wind speed of 59 m/s per minute on average) or higher increased from about 20% in the 1970s to about 35% since the 1990s. Murakami et al. (2012b) conducted a future climate experiment using a high-resolution GCM with a 20-km mesh. They found that the frequency of TCs at the end of the 21st century (under the A1B scenario) will decrease (13–25%), but their intensity will increase by 1–7% globally, although the frequency of TCs will decrease at the end of the 21st century (under the A1B scenario). Knutson et al. (2010) found that under the A1B scenario, the frequency of TCs at the end of the 21st century (under the A1B scenario) would decrease (13–25%), but the intensity would increase by approximately 1–7% globally. Knutson et al. (2010) reported that the intensity of TCs at the end of the 21st century will increase by about 2–11% globally, and precipitation will increase by 20% near the center of TCs. Knutson et al. (2015) conducted a future climate experiment similar to that of Knutson et al. (2010) using the Representive Concentration Pathways (RCP) scenario used in the IPCC Fifth Assessment Report as their setup scenario. The study of future changes in TCs using a numerical model under the RCP 4.5 scenario shows that the global average frequency of TCs decreases (−16% compared to the present climate), but the frequency tends to increase (+24%) when limited to strong TCs (Category 4 or 5), and that the frequency of TCs increases (+20%) when limited to strong TCs (Category 4 or 5). The frequency of TCs with maximum wind speeds exceeding 65 m/s (+59%) has been shown to increase. The increase in TC intensity was also noted to increase by about 4% on average compared to the present climate globally (Knutson et al., 2015). Some studies focused on TCs, especially typhoons.

Tsuboki et al. (2015) conducted a quasi-warming downscaling experiment (SRES A1B scenario) focusing on 30 strong cases of typhoons from 1979 to 1993. The strongest typhoons developed to a minimum central pressure of 857 hPa by the end of the 21st century. Furthermore, Kanada et al. (2017) conducted a pseudo-warming downscaling experiment on Ise Bay typhoon (1959) and noted future changes in structure and intensity at the end of the 21st century (RCP 8.5 scenario). The results of the future climate calculations report an increase in the height and upwelling of the typhoon eye wall cloud and an increase in typhoon intensity.

Finally, in a study focusing on the location and path of TCs in the Atlantic, Murakami and Wang (2010) conducted a future climate experiment at the end of the 21st century under the A1B scenario for TCs in the Atlantic. They revealed that the location of TCs will shift from the Western Atlantic, which is currently the main area of occurrence, to the Northwestern and Eastern Atlantic in the future. As a study focusing on the Northwest Pacific Ocean, Wu and Wang (2004) conducted future climate experiments in the early 21st century (2000–2029) and mid-21st century (2030–2059) (under the A2 and B2 scenarios). The results indicate that in the Northwest Pacific, the path of typhoons shifts to the south in the early 21st century and to the north in the mid-21st century owing to changes in the directional currents that affect typhoon paths. Yamaguchi et al. (2020) evaluated future changes in TC (typhoon) movement speeds using a large number of numerical simulation results. They found that typhoons moving through the mid latitudes where Japan is located will slow down by approximately 10% by the end of the 21st century, assuming no mitigation measures are taken. The abovementioned studies have shown that the frequency of TCs tends to decrease, while the frequency of strong TCs and intensity of TCs tend to increase, and the areas where they occur, their paths and their movement speeds change with climate change owing to global warming. The increase in the intensity of TCs as an external force is also expected to increase storm surges. Many studies have been reported using regional climate models (RCMs) to evaluate the impact of changes in typhoon intensity on storm surge (e.g., Takayabu et al., 2015; Mori and Takemi, 2016). Takayabu et al. (2015) compared the warming impact of Typhoon Haiyan between present and past climates. Mori and Takemi (2016) used the domain model WRF (Weather Research and Forecasting model) and a pseudo-warming downscaling method to compare Typhoons Vera (1959) and Haiyan using the domain model WRF and the pseudo-warming downscaling method. They conducted a future climate experiment at the end of the 21st century under the RCP 8.5 scenario. The results show that storm surges increased by approximately 1.0–1.3 m with increasing typhoon intensity (mean future change in peak intensity is −14.1 hPa), especially for Typhoon Ise Bay. Thus, storm surges caused by typhoons tend to increase because of global warming.

1.5 RECENT ADVANCES IN TYPHOON FORECASTING RESEARCH

As mentioned above, typhoons have caused significant damage in the Pacific and Atlantic coastal areas. Although knowledge on typhoons is still being accumulated, it is important to improve the accuracy of typhoon forecasting in particular, as it is fundamental information for disaster prevention. Ito (2023) reported on the recent technology for typhoon intensity forecasting. In his report, he states that in addition to numerical and statistical modeling, the importance of direct observation by airplanes is being recognized again. For example, in the North Atlantic and Northeast Pacific, the US military and National Oceanic and Atmospheric Administration (NOAA) continue direct observations by aircraft on a commercial basis. As an example, the US military conducts penetration flights into the eye of typhoons using the WC-130J aircraft. The aircraft is equipped with a dropsonde, a probe to acquire flight-level data and an instrument called Stepped Frequency Microwave Radiometer (SFMR), which uses microwaves to observe wind speed through remote sensing. WP-3D, NOAA's aircraft, is capable of penetrating typhoons and can observe wind speeds with a radar called the tail Doppler radar in addition to the usual instruments. Furthermore, aircraft observations for research purposes are sometimes conducted by other organizations. In the Northwest Pacific Ocean, the Dropsonde Observation for Typhoon Surveillance near the TAiwan Region (DOTSTAR) observation by Taiwan and the Hong Kong observation are aircraft observations that have been conducted permanently since the end of the typhoon aircraft observation in 1987. Additionally, several other projects have included aircraft observations, but direct observations of conditions near the center of strong typhoons have been limited to a few cases, such as THORPEX Pacific Asian Regional Campaign (T-PARC) in 2008 and Impact of Typhoons on the Ocean in the Pacific (ITOP) in 2010. Therefore, Professor Kazuhisa Tsuboki of Nagoya University has taken the lead in implementing T-PARCII, the first large-scale project led by a Japanese researcher that includes aircraft observations of typhoons. In this project, a dropsonde, which is dropped from an aircraft and transmits wind speed, temperature, humidity and pressure to the aircraft, was developed in FY2016, and a penetration flight and observation of Typhoon No. 21 in 2017 were successfully conducted. A small jet aircraft, Gulfstream-II, was used for this observation, and the penetrating flight was achieved by flying at a very high altitude of 43,000 ft (Fudeyasu et al., 2018). On 21 October 2017, the first day of observation, the central pressure was 920 hPa, and wind speeds of 80 m/s at an altitude of 1.3 km near the wall cloud and 58 m/s at an altitude of 14.2 m were observed. JMA's estimate of the central pressure of the typhoon was 930 hPa, which differed from the actual measurement, and the warm air core above the center of the typhoon had a dual structure (Ito et al., 2018; Yamada et al., 2021). The T-PARCII project has also succeeded in direct observation of Typhoon No. 24 in 2018, Typhoon No. 16 in 2021 and Typhoon No. 14 in 2022. Particularly, the rapid decay of Typhoon No. 24 in 2018 (Hirano et al., 2022) and the rapid development of Typhoon No. 14 in 2022 were captured, and future research progress is expected. In addition to direct observation by aircrafts, a method to estimate wind speed using Doppler reflection with ground-based radar has been developed. Generally, a single radar can only measure wind speed in the line-of-sight direction, but assuming that typhoons have a center and that concentric tangential wind speeds prevail inside the center, an arial wind speed distribution can be estimated with only one radar (Lee et al., 1999). Convection in wall clouds is important for typhoon intensity forecasting. If this is to be reproduced explicitly in a high-resolution model, the grid spacing should be less than a few kilometers and a coupled atmosphere–ocean model should be used. On the other hand, as typhoon intensity is constrained by the vertical shear of the environmental field and ocean heat storage on a scale of several hundred kilometers, a semi-statistical model that combines the output of a low-resolution model with ocean and other information can be used to approximate typhoon intensity forecasts without fully reproducing typhoon intensity on a high-resolution numerical model. JMA began to make full-scale use of semi-statistical models, such as Typhoon Intensity Forecasting Scheme (TIFS), in 2017; the errors in strength forecasts released to the public decreased by approximately 20% compared to that in mid-2000s to mid-2010s. In addition, since 2019, the JMA has been conducting five-day forecasts of typhoon intensity, and other efforts in current forecasting are also becoming more active. Forecasting the rapid development of typhoons is difficult and challenging. However, it is expected to improve in the future through the effective use of nonlinear statistical models and satellites. Recently, there has been progress in the use of artificial intelligence (AI), and it can be said that the technology for estimating typhoon intensity is steadily advancing.

"Prospects for Future Typhoon Prediction Research," published by the Meteorological Society of Japan (Sato et al., 2022), introduces the items necessary to make progress in improving the accuracy of typhoon prediction by the year 2050. In this report, the following points are introduced:

- Further development of scientific understanding of typhoons.

- Realization of more dense and frequent observations.

- Development of more accurate numerical models that can realistically represent typhoons.

- Development of data assimilation methods that optimally combine observations and numerical models.

- Development of higher-performance supercomputers.

Many research issues are listed in this chapter about preparation for typhoon attacks, which are expected to become more severe with the progression of global warming in the future. This is the summary of the report. As described above, the future typhoon forecasting technology is expected to improve and storm surge forecasting will also be advanced by utilizing the advances in computers, observation accuracy and AI without any deficiencies.

1.6 CONTENTS OF THIS BOOK

Storm surges associated with TCs have been causing damage in many parts of the world. Recently, they have shown a tendency to increase in severity. However, these calculations are not realistic except for those conducted by relatively large organizations, such as national institutions and research institutes, because of the huge computational costs involved. In fact, in Japan and other typhoon-prone countries, empirical estimation is still an active method used by local administrative agencies (prefectures and cities) and practitioners for countermeasures. This book describes the estimation methods of TC weather fields that are widely used in practice (Chapter 2), types of storm surge estimation models (Chapter 3), characteristics and limitations of empirical methods and future perspectives (Chapter 4). Chapter 5 presents future prospects for empirical typhoon model (ETM) storm surge forecast and a summary of this book.

REFERENCES

Cabinet Office, Government of Japan (2008). Cyclone damage in Myanmar. Overseas Disaster Report, Cabinet Office, Government of Japan, http://www .bousai.go.jp/kohou/kouhoubousai/h20/07/repo_02.html (accessed March 27, 2024). (in Japanese).

Christensen, J. H., & Co authors (2013). Climate phenomena and their relevance for future regional climate change. In T. F. Stocker et al. (Eds.), *Climate change 2013: The physical science basis*. Cambridge University Press, pp. 1217–1308.

Fire and Disaster Management Agency (2019). Damage caused by Typhoon No. 21 in 2008 and the status of response by firefighting organizations, etc. (10th report) https://www.fdma.go.jp/disaster/info/items/40fa100bdc7b7db0e896 733faa88c208d8b032ee.pdf (accessed March 27, 2024). (in Japanese).

Fudeyasu, H., Yamada, H., Miyamoto, Y., Ito, K., Yamaguchi, M., & Kanada, S. (2018). *What we know and don't know about typhoons*. Beret Publishing.

Hirano, S., Ito, K., Yamada, H., Tsujino, S., Tsuboki, K., & Wu, C.-C. (2022). Deep eye clouds in tropical cyclone trami (2018) during T-PARCII dropsonde observations. *Journal of the Atmospheric Sciences*, 79, 683–703.

IPCC (2021). Summary for policymakers. In: Climate Change 2021: the physical science basis. contribution of working group I to the sixth assessment report of the Intergovernmental Panel on Climate Change [Masson-Delmotte, V, P. Zhai, A. Pirani, S.L. Connors, C.Péan, S. Berger, N. Caud, Y. Chen, L. Goldfarb, M. I. Gomis, M. Huang, K. Leitzell, E. Lonnoy, J.B.R. Matthews, T. K. Maycock, T.Waterfield, O. Yelekçi, R. Yu, B. Zhou (eds.)]. In Press, 40p

Ito, K. (2023). Recent progresses in the research on tropical cyclone intensity. *Wind Engineers. JAWE*, 48(3), 261–267.

Ito, K., Yamada, H., Yamaguchi, M., Nakazawa, T., Nagahama, N., Shimizu, K., Ohigashi, T., Shinoda, T., & Tsuboki, K. (2018). Analysis and forecast using dropsonde data from the inner-core region of Tropical Cyclone Lan (2017) obtained during the first aircraft missions of T-PARCII. *SOLA*, 14, 105–110.

Japan Meteorological Agency (2013). Typhoon 1330 location chart. https://www .data.jma.go.jp/fcd/yoho/data/typhoon/T1330.pdf (accessed March 27, 2024).

Japan Meteorological Agency (2018a). Typhoon 1821 location chart. https:// www.data.jma.go.jp/fcd/yoho/data/typhoon/T1821.pdf (accessed March 27, 2024). (in Japanese).

Japan Meteorological Agency (2018b). Wind storm, storm surge, etc. due to Typhoon No. 21 in 2018, 41p, https://www.data.jma.go.jp/obd/stats/data/ bosai/report/2018/20180911/20180911.html (accessed March 27, 2024).

Japan Meteorological Agency (2023). Examples of weather causing disasters (1989–2023). https://www.data.jma.go.jp/obd/stats/data/bosai/report/index_1989 .html (accessed March 27, 2024). (in Japanese).

Japan Society of Civil Engineers, Coastal Engineering Committee (2018). Coastal disaster caused by Typhoon Jebi in 2018. http://www.coastal.jp/ (accessed March 27, 2024). (in Japanese).

Kanada, S., Takemi, T., Kato, M., Yamasaki, S., Fudeyasu, H., Tsuboki, K., Arakawa, O., & Takayabu, I. (2017). A multimodel Intercomparison of an intense typhoon in future warmer climates by four 5-km-mesh models. *Journal of Climate*, 30, 6017–6036.

Knutson, T. R., McBride, J. L., Chan, J., Emanuel, K., Holland, G., Landsea, C., Held, I., Kossin, J. P. Srivastava. A. K., & Sugi, M. (2010). Tropical cyclones and climate change. *Nature Geoscience*, 3, 157–163.

Knutson, T. R., Sirutis, J., Zhao, M., Tuleya, R., Bender, M., Vecchi, G., Villarini, G., & Chavas, D. (2015). Global projections of intense tropical cyclone activity for the late twenty-first century from dynamical downscaling of CMIP5/RCP4.5 scenarios. *Journal of Climate*, 28, 7203–7224.

Lander, M., Guard, C., & Camargo, S. J. (2014). Super Typhoon Haiyan. *State of the Climate in 2013*, pp. S112–S113.

Lee, W.-C., Jou, B. J.-D., Chang, P.-L., & Deng, S.-M. (1999). Tropical cyclone kinematic structure retrieved from single Doppler radar observations. Part I: Interpretation of Doppler velocity patterns and the GBVTD technique. *Monthly Weather Review*, 127, 2419–2439.

Lin, I.-I., Pun, I.-F., & Lien, C.-C. (2014). "Category-6" supertyphoon Haiyan in global warming hiatus: Contribution from subsurface ocean warming. *Geophysical Research Letters*, 41, 8547–8553.

Mori, N., & Takemi, T. (2016). Impact assessment of coastal hazards due to future changes of tropical cyclones in the North Pacific Ocean. *Weather and Climate Extremes*, 11, 53–69.

Mori, N., Yasuda, T., Arikawa, T., Kataoka, S., Nakajo, K., Suzuki, Y., Yamanaka, Y., & Webb, A. (2019). 2018 Typhoon Jebi post-event survey of coastal damage in the Kansai region. *Japan. Coastal Engineering Journal*, 61(3), 278–294.

Murakami, H., & Wang, B. (2010). Future change of north Atlantic tropical cyclone tracks: projection by a 20-km-mesh global atmospheric model. *Journal of Climate*, 23, 2699–2721.

Murakami, H., Mizuta, R., & Shindo, E. (2012a). Future changes in tropical cyclone activity projected by multi-physics and multi-SST ensemble experiments using the 60-km-mesh MRI-AGCM. *Climate Dynamics*, 39, 2569–2584.

Murakami, H., Wang, Y., Yoshimura, H., Mizuta, R., Sugi, M., Shindo, E., Adachi, Y., Yukimoto, S., Hosaka, M., Kusunoki, S., Ose, T., & Kitoh, A. (2012b). Future changes in tropical cyclone activity projected by the new high-resolution MRI-AGCM. *Journal of Climate*, 25(9), 3237–3260.

National Institute for Land and Infrastructure Management (2005). Hurricane Katrina Disaster Investigation Report, 8p, https://www.ysk.nilim.go.jp/kakubu/engan/engan/pdf/hariken20051101h.pdf (accessed March 27, 2024). (in Japanese).

National Research Institute for Earth Science and Disaster Resilience (2006). Characteristics of the 2005 U.S. Hurricane Katrina Disaster, 22p, https://dil-opac.bosai.go.jp/publication/nied_natural_disaster/pdf/41/41-01.pdf (accessed March 27, 2024). (in Japanese).

Oouchi, K., Yoshimura, J., Yoshimura, H., Mizuta, R., Kusunoki, S., & Noda, A. (2006). Tropical cyclone climatology in a global-warming climate as simulated in a 20 km-mesh global atmospheric model frequency and wind intensity analysis. *Journal of the Meteorological Society of Japan*, 84(2), 259–276.

Osaka Large-Scale Urban Flooding Countermeasures Study Group (2018). Osaka large-scale urban flooding countermeasures guidelines, 214p. https://www .kkr.mlit.go.jp/bousai/sonae/oosakadaikibo/ol9a8v000000leyk-att/guideline-. pdf (accessed March 27, 2024). (in Japanese).

Sato, M. Y., Hisashi, K., Ito, H., Fudeyasu, T., Miyoshi, T., Kawabata, K., Tsuboki, Horinouchi, T., Okamoto, K., Yamaguchi, M., Nakano, M., Wada, A., Kanada, S., & Tsujino, S. (2022). Prospects for future typhoon forecasting research. *Tenki*, 65(5), 41–50.

Tajima, Y., Yasuda, T., Pacheco, B. M., Cruz, E. C., Kawasaki, K., Nobuoka, H., Miyamoto, M., Asano, Y., Arikawa, T., Ortigas, N.M., Aquino, R., Mata, W., Valdez, J., & Briones, F. (2014). Initial report of JSCE-PICE joint survey on the storm surge disaster caused by Typhoon Haiyan. *Coastal Engineering Journal*, 56(01), 65–76.

Takayabu, I., Hibino, K., Sasaki, H., Shiogama, H., Mori, N., Shibutani, Y., & Takemi, T. (2015). Climate change effects on the worst-case storm surge: a case study of typhoon Haiyan. *Environmental Research Letter*, 10, 064011.

The Asahi Shimbun (2018). September 6, 2018 Article.

The General Insurance Association of Japan (2019). End-of-year survey of the number of claims paid, claims paid (including estimates), etc., for various types of property insurance for windstorms occurring in fiscal 2018. https:// www.sonpo.or.jp/news/release/2019/1905_02.html (accessed March 27, 2024). (in Japanese).

Toyoda, M., Yoshino, J., Arakawa, S., & Kobayashi, T. (2015). Numerical experiments of typhoon Haiyan (2013) and its storm surge using a high-resolution coupled typhoon-ocean model. *Journal of Japan Society of Civil Engineers, Ser. B2 (Coastal Engineering)*, 71(2), I_463–I_468. (in Japanese).

Tsuboki, K., Yoshioka, M. K., Shinoda, T., Kato, M., Kanada, S., & Kitoh, A. (2015). Future increase of supertyphoon intensity associated with climate change. *Geophysical Research Letters*, 42, 646–652.

Webster, P. J., Holland, G. J., Curry, J. A., & Chang, H.-R. (2005). Changes in tropical cyclone number, duration, and intensity in a warming environment. *Science*, 309, 1844–1846.

World Meteorological Organization (2022). State of the global climate, 55p. Retrieved June 25, 2024 from https://library.wmo.int/records/item/66214-state-of-the-global-climate-2022

Wu, L., & Wang, B. (2004). Assessing impact of global warming on tropical cyclone tracks. *Journal of Climate*, 17, 1686–1698.

Yamada, H., Ito, K., Tsuboki, K., Shinoda, T., Ohigashi, T., Yamaguchi, M., Nakazawa, T., Nagahama, N., & Shimizu, K. (2021). The double warm-core structure of Typhoon Lan (2017) as observed through the first Japanese eyewall penetrating aircraft reconnaissance. *Journal of the Meteorological Society of Japan*, 99, 1297–1328.

Yamaguchi, M., Chan, J. C. L., Moon, I. J., Yoshida, K., & Mizuta, R. (2020). Global warming changes tropical cyclone translation speed. *Nature Communications*, 11, 47.

Yoshimura, J., Sugi, M., & Noda, A. (2006). Influence of greenhouse warming on tropical cyclone frequency. *Journal of the Meteorological Society of Japan*, 84(2), 405–428.

CHAPTER 2

Current status of empirical typhoon models for typhoon meteorological fields

Accurate estimation of pressure drops and strong winds associated with tropical cyclones (TCs) can improve the accuracy of coastal disaster prediction related to TCs. Numerical simulations using an regional climate model (RCM) or a coupled atmosphere–ocean coupled model, which can reproduce typhoons, associated waves and storm surges, are being used owing to improved computational capabilities (Mori et al., 2014). Mori et al. (2014) performed numerical experiments for Typhoon Haiyan in 2013 to investigate the characteristics of local storm surge enhancement in the Gulf of Leyte and Tacloban regions of the Philippines. In this study, two 1-km-resolution meteorological models (Weather Research and Forecasting model (WRF); Skamarock et al., 2008, the Cloud Resolving Storm Simulator; Tsuboki and Sakakibara, 2002) and a storm surge model (SSM) were used. Additionally, they reported that the seiche of the Leyte Gulf enhanced the storm surge, and the storm surge in the Leyte Gulf could potentially be amplified by a particular typhoon track.

DOI: 10.1201/9781003478782-2

The use of atmosphere–ocean coupled models for typhoon-induced storm surges is becoming prominent. However, hazard maps, countermeasures and many typhoon disaster assessments for storm surges are widely used at the policy level in the parameterized empirical typhoon model (ETM) and SSM in Japan (Ministry of Land, Infrastructure, Transportation, and Tourism, Japan, 2020). Empirical methods for estimating typhoon meteorological fields have been proposed in many studies (Myers, 1954; Schloemer, 1954; Holland, 1980; Fujii and Mitsuta, 1986; Chavas et al., 2015; Wang et al., 2015). Fujii and Mitsuta (1986) and Chavas et al. (2015) estimated the wind speed distribution. Holland (1980) and Wang et al. (2015) proposed estimation methods for pressure and wind speed distributions. The methods proposed by Schloemer (1954) and Holland (1980) are widely used because of the small number of input variables and simplicity of the equations. Although various other ETMs have been proposed by many researchers (Jelesnianski et al., 1992, Willoughby et al., 2006), the ETM proposed by Schloemer (1954) and used by Holland (1980), Holland et al. (2010) and Fujii and Mitsuta (1986) is explained in this section. ETM is extensively used as a general method in coastal engineering in Japan. Moreover, owing to the low computation cost and ease of handling, ETM has been widely employed for estimating meteorological fields using simplified typhoon information (e.g., minimum central pressure) for storm surges. Therefore, engineers and policymakers are likely to adopt ETM.

2.1 SCHLOEMER TYPHOON MODEL

This meteorological field estimation equation is the oldest of the widely used methods used today. The equation describing the pressure distribution is expressed as follows:

$$P(r) = P_c + \Delta P \exp\left(-\frac{1}{x}\right), \tag{2.1}$$

where $P(r)$ [hPa] is the pressure at a point r km from the typhoon center, P_c [hPa] is the central pressure of the typhoon and ΔP [hPa] is the pressure depression from the atmospheric environmental pressure (here, we used $1013\,\text{hPa} - P_c$); x is expressed as the ratio between the distance r [km] and the radius of maximum wind speed R_w [km] $(x = r / R_w)$. The wind generated by a typhoon is expressed as a composite of the tilt wind caused by the pressure gradient and field wind component generated by the movement of the typhoon.

The expression for the wind speed of the tilt wind is generally given by the following equation.

$$V(r) = -\left(\frac{rf}{2}\right) + \sqrt{\left(\frac{rf}{2}\right)^2 + \frac{\Delta P}{\rho_a}\frac{\partial P(r)}{\partial r}}, \tag{2.2}$$

where $V(r)$ [m/s] is the wind speed, ρ_a [g/m³] is the atmospheric density, f [rad/s] is the Coriolis parameter ($2\times7.29\times10^{-5}\times\sin\varphi$) and φ [rad] is the latitude. Substituting Eq. (2.1) into this equation yields Eq. (2.3):

$$V(r) = -\left(\frac{rf}{2}\right) + \sqrt{\left(\frac{rf}{2}\right)^2 + \frac{\Delta P}{\rho_a}\frac{1}{x}\exp\left(-\frac{1}{x}\right)}. \tag{2.3}$$

Here, the wind speed is assumed to be under ideal conditions with no surface friction. When surface friction is considered, the wind speed reduces, and the wind direction is slightly toward the center of the typhoon. Therefore, the following coefficients are generally used to calculate the surface wind speed V over land.

$$V_1 = C_1 V(r), \tag{2.4}$$

where C_1 is the coefficient of friction with the ground surface (0.6–0.7 is common). The wind component V_2 of the field owing to the movement of the typhoon is given by:

$$V_2 = C_2\frac{V(r)}{V(r_0)}MS, \tag{2.5}$$

where MS is the speed of movement of the TC [m/s] and C_2 is the correction factor (usually 0.6–0.7). The wind direction is assumed to be the same as the direction of movement of the TC. Eqs. (2.1–2.5) are used to estimate the pressure distribution and wind speed at the TC and to set the input meteorological field for storm surge estimation in the target bay. This method has been used in the Northwest Pacific region, especially in Japan, because it requires very little information and only uses readily available information. Even the most recent version (June 2023) of the Japanese Storm Surge Estimation Guide contains the results of this method.

2.2 ETM PROPOSED BY HOLLAND

The empirical model by Holland (1980), which is widely used in North America and other parts of the world, is explained here. The pressure and wind speed distributions are expressed by the following two equations.

$$P(r) = P_c + \Delta P \exp\left(-\frac{1}{x^B}\right) \tag{2.6}$$

$$V(r) = -\left(\frac{rf}{2}\right) + \sqrt{\left(\frac{rf}{2}\right)^2 + \frac{\Delta P}{\rho_a} \frac{B}{x^B} \exp\left(-\frac{1}{x^B}\right)} \tag{2.7}$$

The form of this equation is similar to that provided by Schloemer (1954) with the addition of the scaling parameter B. This scaling parameter is set according to the shape of the TC and is estimated using the following equation.

$$B = \rho_a e V_{max}^2 / \Delta P, \tag{2.8}$$

where e is the base of natural logarithms. This equation assumes the gradient–wind equilibrium and considers the surface friction and moving speed of the TC. This model improves on Schloemer (1954) by using the scaling parameter B, allowing estimation that considers the structure of individual TCs. As an improved form of this equation, Holland et al. (2010) proposed the following equation for wind speed distribution.

$$V_s = \left[\frac{100 B_s \Delta P_s \left(\frac{1}{x}\right)^{B_s}}{\rho_s e^{\left(\frac{1}{x}\right)^{B_s}}}\right]^x, \tag{2.9}$$

where the subscript s refers to surface values (at a nominal height of 10 m). Parameter B_s is related to the original B by $B_s = B g_s^x$, where g_s is the reduction factor for gradient-to-surface winds. If we set $x = 0.5$ and keep B_s constant, as per Holland (1980), then it is often impossible to accurately reproduce the core and external wind structure in hurricanes. Allowing x and/or B_s to vary with the radius substantially improves this reproduction. We experimented with various combinations of varying x and B_s and settled on keeping B_s constant and allowing x to vary linearly. Hence, application of Eq. (2.9) requires the following data: the central pressure (with or without maximum winds), radius of maximum winds, an external pressure and collocated surface wind speed and surface air density.

If the maximum surface winds and central pressure have been directly or reliably and independently observed, then we can estimate the B_s parameter (following Holland, 1980). However, if the analysis uses Dvorak or other remote assessment, we recommend the approach adopted by Holland (2008) for estimating the central pressure first and then deriving the B_s parameter and V_{max}:

$$B_s = -4.4 \times 10^{-5} \Delta P_s^2 + 0.01 \Delta P_s + 0.03 \frac{\partial P_{cs}}{\partial t} - 0.014 \varphi + 0.15 MS^x + 1.0,$$

$$x = 0.6\left(1 - \frac{\Delta P_s}{215}\right),$$

$$V_{max} = \left(\frac{100 B_s}{\rho_{ms} e} \Delta P\right)^{0.5}, \tag{2.10}$$

where $\partial P_{cs}/\partial t$ denotes the intensity changes [hPa/h] and φ is the absolute value of latitude in degrees.

The surface air density can be estimated with good accuracy using Holland's method (2008):

$$\rho_s = \frac{100 P_s}{R T_{vs}},$$

$$T_{vs} = (T_s + 273.15)(1 + 0.61 q_s),$$

$$q_s = RH_s \left(\frac{3.802}{100 P_s}\right) e^{\left(\frac{17.67 T_s}{243.5 + T_s}\right)},$$

$$T_s = SST - 1, \tag{2.11}$$

where $R = 286.9$ J/kg/K is the gas constant for dry air, T_{vs} is the virtual surface temperature (in K), q_s is the surface moisture [g/kg], T_s is the surface temperature, SST is the sea-surface temperature [°C] and RH_s is the surface relative humidity (assumed to be 0.9 in the absence of direct observations). Thus, the only external input required in Eq. (2.11) is an estimate of the SST and the surface pressure from Eq. (2.6). This enables an improved fit with error reduction by several percentage for intense cyclones. However, it is not critical, and a constant value of density can be used if preferred. Alternatively, if there is a good estimate of the maximum winds, then the alternate method in Eq. (2.9) can be used as it does not require explicit estimation of density.

This formula is an improved version of that provided by Holland (1980) and is often used for hurricane estimation. On the other hand, the information required and the complexity of the equation make it a slightly difficult estimation method for practitioners who do not have specialized knowledge.

2.3 ETM PROPOSED BY JELESNIANSKI (1965)

Jelesnianski (1965) proposed a numerical calculation of storm tides, in which the regional difference in pressure distribution is considered. It employs two distinguished systems to describe the field in terms of location. The typhoon wind and pressure field distribution are expressed as Eqs. (2.12) and (2.13), respectively (Gong et al., 2022).

When $0 \leq r \leq R_w$,

$$P_a = P_0 + \frac{1}{4}(P_\infty - P_0)\left(\frac{r}{R_w}\right)^3$$

$$V = \frac{r}{R_w + +r}\left(V_x \boldsymbol{i} + V_y \boldsymbol{j}\right) + V_{max}\left(\frac{R_w}{r}\right)^{\frac{3}{2}}\frac{1}{r}\left(Ai + Bj\right)$$

(2.12)

When $r > R_w$,

$$P_a = P_\infty - \frac{3}{4}(P_\infty - P_0)\left(\frac{R_w}{r}\right)$$

$$V = \frac{R_w}{R_w + +r}\left(V_x \boldsymbol{i} + V_y \boldsymbol{j}\right) + V_{max}\left(\frac{R_w}{r}\right)^{\frac{1}{2}}\frac{1}{r}\left(Ai + Bj\right)$$

, (2.13)

where P_a is the pressure of every grid; P_∞ represents the atmospheric pressure at infinity; P_0 is the pressure of the typhoon center; r is the distance from the center of the TC; R_w is the radius of maximum wind speed; V

is the velocity of every grid; V_{max} is the maximum wind velocity at the typhoon center; V_x and V_y are the movement velocities at the typhoon center along the X- and Y-axes, respectively and i and j are unit vectors on the coordinate axis. $A = -y\cos\theta - x\sin\theta$, $B = x\cos\theta - y\sin\theta$, where x and y are the coordinates of the typhoon center, and θ is the ingress angle. The Jelesnianski typhoon wind model (Eqs. 2.12 and 2.13) defined a typhoon as a circular storm with additional wind velocity. The wind speed V is composed of two components: the circular wind speed and velocity of the storm center. The circular wind speed V' is defined as follows:

$$V' = \begin{cases} V_{max}\left(\dfrac{r}{R_w}\right)^{\frac{3}{2}}\dfrac{1}{r}[Ai + Bj], & 0 \le r \le R_w \\[4mm] V_{max}\left(\dfrac{R_w}{r}\right)^{\frac{1}{2}}\dfrac{1}{r}[Ai + Bj], & r \ge R_w \end{cases} \tag{2.14}$$

This equation can represent the relationship between wind speed and wind radius in a circular storm. This equation can also be transformed as follows:

$$R_w = \begin{cases} C_1 r\left(\dfrac{V'}{V_{max}}\right)^{\frac{-2}{3}}, & 0 \le r \le R_w \\[4mm] C_2 r\left(\dfrac{V'}{V_{max}}\right)^{2}, & r \ge R_w \end{cases} \tag{2.15}$$

where C is the parameter. Then, this equation can be represented as follows:

$$R_w = \alpha r\left(\frac{V'}{V_{max}}\right)^{\beta}, \tag{2.16}$$

where α and β are parameters related to the position and sea condition of the typhoon center and computational grid. The major radius is a relatively easily available data point included in the best track dataset.

2.4 ETM PROPOSED BY FUJII AND MITSUTA

The following equations express the ETM for pressure (Myers, 1954) and wind speed distribution (Fujii and Mitsuta, 1986).

$$P(r) = P_c + \Delta P \exp\left(-\frac{1}{x^B}\right) \tag{2.17}$$

$$V(r) = -\left(\frac{rf}{2}\right) + \sqrt{\left(\frac{rf}{2}\right)^2 + \frac{\Delta P}{\rho_a}\frac{B}{x^B}\exp\left(-\frac{1}{x^B}\right)}, \tag{2.18}$$

where $P(r)$ [hPa] is the pressure at a point r km from the typhoon center, P_c [hPa] is the central pressure of the typhoon and ΔP [hPa] is the pressure depression from the atmospheric environmental pressure (here, we use 1013 hPa $- P_c$); x is expressed as the ratio between the distance r [km] and the radius of maximum wind speed R_w [km] ($x = r / R_w$). In the wind speed equation, $V(r)$ [m/s] is the wind speed, ρ_a [g/m³] is the atmospheric density, f [rad/s] is the Coriolis parameter ($2 \times 7.29 \times 10^{-5} \times \sin\varphi$) and φ [rad] is the latitude. Eq. (2.17) expresses the barometric pressure distribution, and Eq. (2.18) expresses the wind speed distribution as a function of ΔP, r and R_w. Both equations are based on the study conducted by Schloemer (1954). Coefficient B appears in atmospheric pressure and wind speed distributions as a scaling parameter, and $B = 1$ is empirically proposed by Myers (1954) and Fujii and Mitsuta (1986). Furthermore, it is necessary to consider the effect of typhoon movement and friction on the wind speed in Eq. (2.18). Blaton's equation is widely used to evaluate the effects of typhoon movement.

$$\frac{1}{r_t} = \frac{1}{r} + \frac{1}{V(r)}\frac{\partial \psi}{\partial t}, \tag{2.19}$$

where t indicates time, r_t is the radius of curvature of the streamline at t and ψ indicates the particle traveling direction. This equation indicates that stronger winds blow on the right side of the direction of typhoon movement. The wind speed, considering the effect of movement, can be obtained from Eqs. (2.18) and (2.19). Additionally, considering the ground surface friction, the surface wind speed was estimated by multiplying the wind speed reduction coefficient $C(x)$ according to the super gradient–wind method proposed by Fujii and Mitsuta (1986).

$$C(x)=C(\infty)+\left[C(x_p)-C(\infty)\right]\left(\frac{x}{x_p}\right)^{k-1}\times\exp\left\{\left(1-\frac{1}{k}\right)\left[1-\left(\frac{x}{x_p}\right)^k\right]\right\},\,(2.20)$$

where $x=r/R_w$, x_p is the value of x that maximizes $C(x)$, k is the parameter corresponding to the shape of the typhoon and $C(\infty)$ is $C(x)$ at infinity from the typhoon. In Japan, x_p is set to be 0.5, k is set to be 2.5, $C(\infty)$ is set to be 2/3 and $C(x_p)$ is set to be 1.2 empirically. These values are the same as those reported in previous studies (Matoba et al., 2006; Yamaguchi et al., 2012). The current ETM calculates atmospheric pressure using Eq. (2.17) and the wind speed using Eqs. (2.18–2.20). This ETM is widely used for assuming storm surge inundation in Japan.

A well-known problem in ETMs is a large error in pressure and wind speed estimated in an approaching typhoon relative to actual observations (Murakami et al., 2007). This could be attributed to the fixed-parameter B generally adopted for all typhoons. The relational equation is as follows (Holland, 1980):

$$\frac{R_p}{R_w}=\left[\frac{B}{(B+1)}\right]^{\frac{1}{B}},\tag{2.21}$$

where R_p [km] is the distance from the center of the typhoon to the point with the maximum pressure gradient. Eq. (2.21), as shown in Eqs. (2.6 and 2.7), is also based on the equation proposed by Schloemer (1954), which is a relationship proposed by Holland (1980) using a scaling parameter. Eq. (2.21) shows that parameter B is related to R_p and R_w. These factors are not constant and need to be changed according to the structure of the typhoon (Holland, 1980; Hubbert et al., 1991). Eq. (2.21) shows that if the ratio of R_p and R_w (radius ratio; R_p/R_w) can be estimated, it is possible to calculate parameter B backward. However, it is generally difficult to use R_p/R_w because the values of R_p and R_w are not obtained. Estimates of R_w are available from satellites and equations for some of the best tracks. However, R_w is not included in the best track of the Japan Meteorological Agency (JMA). Moreover, no estimation was disclosed for R_p. A few studies have investigated the relationship between TC characteristics. For example, Vickery and Wadhera (2008) analyzed TCs in the Atlantic and found that B has a strong negative correlation with R_w. Additionally, B has a positive correlation with TC intensity. However, they also reported that the estimation equation for B changes depending on the target area. Therefore, it is necessary to have an appropriate relationship of B based on the typhoon characteristics.

2.5 UNCERTAINTY OF RADIUS OF MAXIMUM WIND SPEED (R_w)

As mentioned in Section 2.4, R_w is usually difficult to obtain from forecasts or observations. In the Atlantic and other areas where aircraft observations are available, it is possible to estimate R_w. However, there is little information available in the Northwest Pacific region. On the other hand, R_w is a very important parameter because it is used in all previous ETM equations. In this section, we review studies on the estimation equation of R_w and discuss the uncertainty of R_w in the operation of ETM.

R_w literally means the distance from the center of the typhoon to the point where the maximum wind speed appears. When the typhoon is near its peak and has a relatively concentric structure, R_w does not vary so much and is relatively stable. On the other hand, near landfall and during the decay phase, R_w varies greatly owing to the collapse of structures and influence of land (Moro et al., 2023). Kossin et al. (2007) proposed the following relationship focusing on the maximum wind radius R_w and eye size R_{eye}.

$$R_w = 2.8068 + 0.8361 R_{eye} \qquad (2.22)$$

However, it is difficult to apply the eye size to all regions because it requires information, such as aircraft observations. In contrast, the following equation using the central pressure P_c is often used in coastal engineering in Japan (Kato, 2005).

$$R_w = 80 - 0.769 \left(950 - P_c\right) \qquad (2.23)$$

This estimation formula is applicable to relatively strong typhoons below 950 hPa. For typhoons above 950 hPa, the following equation is proposed.

$$R_w = 80 - 0.769 \left(950 - P_c\right) \qquad (2.24)$$

Kawai et al. (2005) proposed the following as a simple estimation equation.

$$R_w = 94.89 \exp^{(P_c - 967)/61.5} \qquad (2.25)$$

On the other hand, although these equations are very simple, they are all based on the central pressure and cannot fully consider changes in the wind field that are not reflected in the pressure. Therefore, although it is possible to express the average trend of R_w using the central pressure, the average value may deviate significantly if it is actually used for storm surge analysis (Takagi and Wu, 2015). Therefore, Takagi and Wu (2015) proposed the following equation as an estimation equation for typhoons that hit the southern sea area of Japan.

$$R_w = 0.23 R_{25} \tag{2.26}$$

where R_{25} is the storm radius (the radius above 25 m/s). In the case of an elliptical typhoon structure, the average value of the major and minor axes is used. Note that this formula is applicable to typhoons below 935 hPa, which basically means that it is applied to typhoons near their peak. Although relatively accurate results are obtained by narrowing down the typhoon intensity, the variability is still large, and a wide range of analysis is reportedly necessary.

Toyoda et al. (2018a) proposed an estimation equation for R_w using multiple regression analysis by reproducing 52 cases of typhoons that hit Japan during 2000–2017. In this report, two equations are proposed using easily available information published by JMA, with 20° N latitude as the boundary.

$$R_w = -1.17 V_{max} - 0.004 R_{25} + 0.03 R_{15} + 3.54\varphi + 38.1 \qquad \varphi \le 20.0$$
$$R_w = -1.5 V_{max} - 0.09 R_{25} + 0.09 R_{15} + 1.54\varphi + 68.8 \qquad \varphi > 20.0 \tag{2.27}$$

Here, R_{15} is the strong wind radius (the radius above 15 m/s) and φ is the latitude (degree). Although the uncertainty due to the estimation equation is large in this method, it is proposed as a reasonable estimation equation because it is applicable to all typhoons and there are few estimation equations based on information from multiple cases.

As described above, this is a very simple and accurate method, but there are still some problems in parameter setting. Although the method considers the characteristic wind distribution, such as the movement speed and super gradient wind, it does not consider the influence of the weather field surrounding the TC, and there is no representation of precipitation. To solve these limitations, numerical calculations based on dynamic weather models have been widely used in recent years.

2.6 NUMERICAL SIMULATIONS USING DYNAMICAL METEOROLOGICAL MODELS

Many ETMs have been used in TC research and practice since the late 20th century. However, with the development of computational instruments, dynamic weather models are becoming more common at the research level. Iwamoto et al. (2014) developed a meteorological-storm surge-tide coupled model and conducted a hindcast experiment on the storm surge by Typhoon Roke (2011) in Tokyo Bay, Japan. The overall behavior of the storm surge was reproduced with high accuracy using the coupled model, but the peak intensity of the storm surge tended to be underestimated. Many studies have shown the validity of numerical storm surge simulations. Recently, the number of studies on storm surge forecasting using the atmosphere–ocean coupled model, which can consider a more realistic meteorological condition, has increased because of an increase in computing power. As described above, many studies have incorporated dynamic weather models to accurately represent weather fields and to estimate storm surges by receiving pressure and wind information from weather models. WRF, which is considered to be the most widely used model globally among various dynamical weather models, is briefly introduced in this section.

WRF is a model for weather forecasting and academic research developed by the WRF project, a joint project led by the National Center for Atmospheric Research and National Center for Environmental Prediction (NCEP). WRF is a fully compressible nonhydrostatic model. The system of basic equations consists of the equation of motion, equation of continuity, equation of state, conservation equation of heat and conservation equation of mixing ratios of water vapor, cloud water, rain water, snow, ice and hail. These basic equations are transformed into a topographic mass coordinate system (hydrostatic coordinate system of dry atmosphere), and then into a map-projected coordinate system to derive the basic equations as a model. For information on the basic equations, numerical scheme and physical model, please refer to Skamarock et al. (2008). There are 57,800 registered users in more than 160 countries (as of 2021). The authors also conducted storm surge estimation using the WRF weather field (e.g., Toyoda et al., 2023) and successfully reproduced the peak and landfall intensities of Typhoon Trami in 2018 with high accuracy. The bias is −1.04 m/s, the root mean square error (RMSE) is 3.56 m/s and the correlation coefficient (CORR) is 0.78, all of which prove the reproducibility with high accuracy. Recently, WRF has also been increasingly applied to future climate calculations that consider climate change effects (e.g., Nakamura et al., 2020). As described above, WRF is a very useful model at the research level. On the other hand, similar dynamical numerical models are still rarely used by port managers and administrations, although they are used by national organizations and laboratories of large companies. Typical reasons for this are that they are overwhelmingly more computationally expensive than empirical models and require a certain level of expertise to handle owing to the large number of schemes and flexible parameter settings. This is not a problem that will be solved immediately; hence, there is a need to continue operating and updating empirical models for practitioners.

Table 2.1 summarizes the characteristics and differences between ETMs and dynamic weather models. As explained, the empirical model is based on estimating equations for pressure and wind speed distributions, and although the computational cost is low, the main focus is on calculating the main body of the TC. Therefore, it is necessary to be careful when using the model because it is characterized by poor accuracy at points far from the TC. Empirical models, which are currently widely used in the coastal field, do not consider precipitation and can only be used to estimate atmospheric pressure and wind speed. On the contrary, dynamical models solve the meteorological field of interest based on the equations of motion and can compute a realistic 3D meteorological field at a huge computational cost. However, the large amount of input data and the high degree of handling difficulty are major challenges of this model.

A report on the reproducibility of meteorological fields and associated storm surge calculations using ETM is presented here. The cases presented are two patterns: successful and unsuccessful cases using ETM. First, a case study of successful ETM calculations is presented, in which the wind field and storm surge due to Typhoon Jebi in 2018 were reproduced. In this case study, the weather field was calculated using the most basic ETM by Myers, and the storm surge calculation was performed using an SSM with nonlinear long wave equations. By using JMA observations and estimated equations, we succeeded in reproducing the Jebi wind field with good accuracy in terms of time-series variation, although there is a slight tendency to underestimate the wind field. Using this meteorological field, we have also succeeded in reproducing the storm surge that occurred in Osaka Bay. The storm surge was reproduced with high accuracy, even during periods when the wind was dominant, and the maximum storm surge was reproduced with an error of less than 10 cm at the Port of Osaka.

TABLE 2.1   Comparison between empirical and dynamical models

	Empirical model (ETM)	**Dynamical model**
Equations	Empirical formulae	Equation of motion
Computational cost	Low	High
Difficulty	Low-middle	Middle-high
Target	Main body of TC	Meteorological field including TC
Rainfall	OFF	ON
Input	Central pressure, local pressure, R_w, position	Spatio-temporal meteorological data

The behavior of drifting debris during storm surges was also examined. The study reported that the anchor paths and final positions of vessels were generally consistent with the observations. Therefore, this is an example of a successful reproduction of a typhoon weather field without using an expensive dynamic model.

Next, Murakami et al. (2007) reported a pattern that could not be reproduced by ETM. In this case, the reproducibility of offshore wind speeds using ETM by Fujii and Mitsuta was compared with the results of the dynamic weather model MM5 for Typhoon Chaba (2004). The results show that the dynamic model results are consistent with observations. However, the ETM results show large differences in the maximum wind speeds and their onset times, and that the friction velocity, which is directly input as wind stress, shows even larger differences than the wind speed. Therefore, it can be said that the dynamic model should be operated for cases where the sea area is greatly affected by complex topography or for the purpose of calculating storm surge over a wide area.

2.7 ESTIMATING TYPHOON METEOROLOGICAL FIELDS USING ARTIFICIAL INTELLIGENCE (AI)

Finally, as a method that does not belong to either the dynamical weather model or ETM estimation methods, we briefly introduce an AI-based weather field estimation method that has garnered attention recently (Higa et al., 2021; Wang et al., 2022; Mulia et al., 2023).

Wang et al. (2022) designed a deep convolutional neural network (CNN) set to estimate the intensity of TCs over the Northwest Pacific Ocean from the brightness temperature data observed by the Advanced Himawari Imager onboard the geostationary satellite Himawari-8. To train the CNN model, 97 TC cases from 2015 to 2018 were used. The comparison results show that the choice of different infrared channels has a significant impact on the performance of TC intensity estimation from the CNN model. Validation against best track data for maximum wind speeds showed that the best multisegmented CNN classification model combining four channels of data as input resulted in very high accuracy (84.8%) and low RMSE (5.24 m/s) and mean bias (−2.15 m/s). The authors also reported that adding an attention layer after the input layer of the CNN improves the accuracy of the model. They also reported that this estimation method provides stable results even when there is some noise in the data.

Higa et al. (2021) proposed a method to estimate typhoon intensity classes from a single satellite image. The method is not a data-driven approach using only deep learning, which is often used as a conventional method, but a deep learning method that incorporates domain-specific knowledge of meteorology. Specifically, using VGG-16, a model of the Visual Geometric Group, the authors have achieved classification accuracy that significantly exceeds that of previous studies by utilizing preprocessed images that emphasize the distribution of typhoon eyes, eyewalls and clouds.

By comparing t-distributed stochastic nearest neighbor embedding plots of VGG feature maps with the original satellite images, we confirmed that fisheye preprocessing promotes cluster formation, suggesting that the model can successfully extract image features related to typhoon intensity classes. Furthermore, gradient-weighted class activation mapping was applied to highlight the distribution of the eye and the clouds surrounding the eye, which are important regions for intensity classification. The results suggest that the new estimation method qualitatively obtained a perspective similar to that of domain experts.

Mulia et al. (2023) demonstrated the application of deep learning using our newly proposed approach based on generative adversarial networks (GANs) to emulate the atmospheric forcing fields for simulating storm surges efficiently. Thirty-four historical typhoon events from 1981 to 2012 were selected to train the GAN. The Regional Ocean Modeling System was selected to simulate storm surges caused by typhoons. The speed–accuracy trade-off, which is considered the primary objective of this study, typically encountered in SSMing has been addressed. Furthermore, some physical properties associated with typhoons while undergoing an extratropical transition (ET) or landfall were reasonably attained, which was previously difficult to accomplish using conventional parametric models. The proposed method in this study can improve the standard numerical storm surge modeling with forcings from the parametric model and is also equipped with a useful forecasting feature with up to 12 h lead time. Furthermore, this method is reported to be favorable for an operational storm surge forecasting system, particularly when access to high-performance computing is unavailable or is limited. They suggested that to attain such an improvement, the GAN model should be trained using a hybrid numerical weather prediction (NWP) and parametric model. Furthermore, further improvements in the hydrodynamics section, apart from using higher-resolution grid and bathymetric data, can be achieved by simulating the nonlinear interactions between tide, wave and surge using coupled models.

These are a few examples of recent AI-based methods for typhoon intensity estimation, but similar studies, studies focusing on typhoon paths and lifetimes and a wide range of other estimation methods have been developed. The direct storm surge estimation method using AI will be discussed in Chapter 3.

REFERENCES

Chavas, D. R., Lin, N., & Emanuel, K. (2015). A model for the complete radial structure of the tropical cyclone wind field. Part I: Comparison with observed structure. *Journal of the Atmospheric Sciences*, 72(9), 3647–3662.

Fujii, T., & Mitsuta, Y. (1986). Simulation of winds in typhoons by a stochastic model. *Journal of Wind Engineering*, 28, 1–12.

Gong, Y., Dong, S., & Wang, Z. (2022). Development of a coupled genetic algorithm and empirical typhoon wind model and its application. *Ocean Engineering*, 248, 110723.

Higa, M., Tanahara, S., Adachi, Y. et al. (2021). Domain knowledge integration into deep learning for typhoon intensity classification. *Scientific Reports*, 11, 12972. https://doi.org/10.1038/s41598-021-92286-w

Holland, G. J. (1980). An analytic model of the wind and pressure profiles in hurricanes. *Monthly Weather Review*, 108(8), 1212–1218. https://doi.org/10.1175/1520-0493(1980)108<1212:AAMOTW>2.0.CO;2

Holland, G. J. (2008) A revised hurricane pressure–wind model. *Monthly Weather Review*, 136(9), 3432–3445.

Holland, G.J., Belanger, J, I., & Fritz A. (2010). A revised model for radial profiles of hurricane winds. *Monthly Weather Review*, 138(12), 4393–4401.

Hubbert, G. D., Holland, G. J., Leslie, L. M., & Manton, M. J. (1991). A real-time system for forecasting tropical cyclone storm surges. *Weather and Forecasting*, 6, 86–97. https://doi.org/10.1175/1520-0434(1991)006<0086:ARTSFF>2.0.CO;2

Iwamoto, T., Nakamura, R., Oyama, T., Mizukami, R., & Shibayam, T. (2014). Prediction of storm surge at Tokyo Bay under RCP8.5 scenario by using Meteorological-Surge-Tide coupled model. *Journal of Japan Society of Civil Engineers, Series B2 (Coastal Engineering)* 70(2): 1261–1265. doi:10.2208/kaigan.70.I_1261.

Jelesnianski, C. P. (1965). A numerical computation of storm tides induced by a tropical storm impinging on a continental shelf. *Monthly Weather Review*, 93(16), 343–358.

Jelesnianski, C., Chen, J., & Shaffer, W. (1992). SLOSH: Sea, lake, and overland surges from hurricanes. In *NOAA Technical Report NWS 48, United States Department of Commerce*, NOAA, NWS, Silver Spring, MD (Vol. 48, Issue April).

Kato, F. (2005). Technical note of National Institute for Land and Infrastructure Management, No.275.

Kawai, H., Honda, K., Tomita, T., & Kakinuma, T. (2005). Technical note of the Port and Airport Research Institute, No. 1103.

Kossin, J. P., Knaff, J. A., Berger, H. I., Herndon, D. C., Cram, T. A., Velden, C. S., Murnane, R. J., & Hawkins, J. D. (2007). Estimating hurricane wind structure in the absence of aircraft reconnaissance. *Weather Forecast*, 22, 89–101.

Matoba, M., Murakami, K., & Shibaki, H. (2006). Super gradient wind (SGW) based typhoon wind estimation and storm surge simulation. *Journal of Japan Society of Civil Engineers, Series B2 (Coastal Engineering)*, 53, 206–210. https://doi.org/10.2208/proce1989.53.206

Ministry of Land, Infrastructure, Transport and Tourism, Japan (2020). Storm surge inundation assumption area figure making manual (ver.2.00), 84pp.

Mori, N., Kato, M., Kim, S., Mase, H., Shibutani, Y., Takemi, T., Tsuboki, K., & Yasuda, T. (2014). Local amplification of storm surge by Super Typhoon Haiyan in Leyte Gulf. *Geophysical Research Letters*, 41(14), 5106–5113.

Moro, Y., Toyoda, M., Kato, S., & Yoshino, J. (2023). Analysis of the trend and variation factors for the radius of maximum wind speed of typhoons making landfall in Japan. *Journal of Japan Society of Civil Engineers, Series B2 (Coastal Engineering)*, 79(17), 23–17165.

Mulia, I. E., Ueda, N., Miyoshi, T. *et al.* (2023) A novel deep learning approach for typhoon-induced storm surge modeling through efficient emulation of wind and pressure fields. *Scientific Reports*, 13, 7918. https://doi.org/10.1038/s41598-023-35093-9

Murakami, T., Yasuda, T., & Yoshino, J. (2007). Reproduction of storm surge generated by typhoon 0416 over a large area using meteorological model and multi-sigma coordinate ocean model. *Journal of Japan Society of Civil Engineers, Series B2 (Coastal Engineering)*, 63(4), 282–290. https://doi.org/10.2208/jscejb.63.282

Myers, V. A. (1954). Characteristics of United States hurricanes pertinent to levee design for lake Okeechobee, Florida. Hydrometeorological Report, U.S. Weather Bureau, No. 32.

Nakamura, R., Kim, K.O., Kato, S., Nakamura, R., & Okabe, T. (2020). Hindcast of Typhoon Jongdari and non-tidal sea level residuals in Mikawa Bay using FDDA in WRF. *Proceedings of the 22nd IAHR-APD Congress 2020, Sapporo, Japan.* 8p.

Schloemer, R.W. (1954). Analysis and synthesis of hurricane wind patterns over Lake Okeechobee, Florida. Hydrometeorological Report, No. 31, 49 pp.

Skamarock, W. C., Klemp, J. B., Dudhia, J., Gill, D. O., Barker, D. M., Duda, M. G., Huang, X-Y., Wang, W., & Powers, J. G. (2008). A description of the advanced research WRF version 3. NCAR Teck. Note 475 113 National Center for Atmospheric Research, Boulder, Colo.

Takagi, H., & Wu, W. (2015). Maximum wind speed radius of typhoon passing through Japanese southern ocean basin, *Journal of Japan Society of Civil Engineers, Series B3*, 71(1), 1–6.

Toyoda, M., Yoshino, J., & Kobayashi, T. (2018a). Estimating empirical formulas of the radius of maximum wind speed in typhoon based on the High-resolution typhoon model *Proceedings of National Symposium on Wind Engineering*, pp.79–84.

Toyoda M., Mori N., Kim SY., Shibutani Y. (2022). Compound flood modeling for small and medium-sized rivers using an integrated model of atmospheric , ocean and river, *Journal of Japan Society of Civil Engineers Series B2* (*Coastal Engineering*), Vol.78, No.2, pp. I_193-I_198.

Tsuboki, K., & Sakakibara, A. (2002). Large-scale parallel computing of cloud resolving storm simulator. In: H.P. Zima, K. Joe, M. Sato, Y. Seo, & M. Shimasaki (Eds.), *High Performance Computing. ISHPC 2002*. Lecture Notes in Computer Science, vol. 2327. https://doi.org/10.1007/3-540-47847-7_21

Vickery, P. J., & Wadhera, D. (2008). Statistical models of Holland pressure profile parameter and radius to maximum winds of hurricanes from flight-level pressure and H*Wind data. *Journal of Applied Meteorology and Climatology*, 47(10), 2497–2517. https://doi.org/10.1175/2008JAMC1837.1

Wang, S., Toumi, R., Czaja, A., & Kan, A. V. (2015). An analytic model of tropical cyclone wind profiles. *Quarterly Journal of the Royal Meteorological Society*, 141(693), 3018–3029.

Wang, C., Zheng, G., Li, X., Xu, Q., Liu, B., & Zhang, J. (2022). Tropical cyclone intensity estimation from geostationary satellite imagery using deep convolutional neural networks. In *IEEE Transactions on Geoscience and Remote Sensing*, vol. 60, pp. 1–16, Art no. 4101416, https://doi.org/10.1109/TGRS.2021.3066299

Willoughby, H. E., Darling, R. W. R., & Rahn, M. E. (2006). Parametric representation of the primary hurricane vortex. Part II: A new family of sectionally continuous profiles. *Monthly Weather Review*, 134(4), 1102–1120. https://doi.org/10.1175/MWR3106.1

Yamaguchi, M., Nonaka, H., Hino, M., & Hatada, Y. (2012). Investigating occurrence possibility of abnormally strong wind estimated using width records of resin freckle left along annual rings of YAKUSUGI. *Journal of Japan Society of Civil Engineers, Series B1 (Hydraulic Engineering)*, 68(4), I-1675–I-1680. https://doi.org/10.2208/jscejhe.68.I_1675

CHAPTER 3

Storm surge models

There are two methods for estimating the magnitude of a storm surge, including storm surge forecasting, and determining the planned storm surge used in designing seawalls, using a formula to estimate the maximum tide-level anomaly based on local storm surge observations. Previously, the estimation method was the most common method from the viewpoint of computational load, but nowadays the numerical method is the most common method owing to the improvement in computer performance.

3.1 STATISTICAL MODEL BY THE JAPAN METEOROLOGICAL AGENCY

The Japan Meteorological Agency (JMA) used an equation relating the maximum tide-level anomaly to the meteorological conditions at the time (maximum wind speed at 10-m height, wind direction, minimum pressure, etc.) based on past storm surge observation for a simple storm surge forecast shown in Eq. (3.1).

$$\zeta_{max} = a\left(1010 - P_{min}\right) + bW^2 \cos\theta + c, \tag{3.1}$$

where ζ_{max} is the maximum tide-level deviation [cm], the term of $a\left(1010 - P_{min}\right)$ is the amount of water-level rise due to the suction effect owing to the drop in air pressure (pressure-driven surge), P_{min} is the minimum central pressure [hPa], the term $bW^2 \cos\theta$ is the amount of water-level rise because of the wind blowing effect on the sea surface (wind-driven surge), W is the maximum wind speed at 10-m height [m/s], θ is the angle

DOI: 10.1201/9781003478782-3

between the main wind direction and wind direction at maximum wind speed W. The coefficients a, b and c of each term are variables that vary with each tide observation station. The values obtained by JMA for major tide stations in Japan are shown in Table 3.1. The value of c is zero, except for places lying on the Sea of Japan side, such as Tonoura, Sakai and Miyazu.

3.2 DYNAMIC STORM SURGE MODEL (SSM)

3.2.1 Nonlinear shallow water equations (NSWEs)

The simplified storm surge estimation formula introduced in Section 3.1 is a simple and useful method, but it can only be used for specific locations where observation data are available for many years. To analyze storm surge in detail, a numerical model (referred to here as a dynamic SSM) using the Navier–Stokes equation (N-S equation) or the NSWEs obtained by approximating N-S equations to a long wave as the governing equation is often used to calculate storm surge anomaly. NSWE, which is composed of a water mass conservation law (continuity equation) and momentum conservation law (equation of motion), calculates the storm surge anomaly and current velocity using the sea-level pressure and wind speed at the sea surface as external forcings. The momentum conservation law determines the vertical distribution of the current velocity so that the surface drift current and bottom friction are in balance. In contrast, the Euler equation under the hydrostatic assumption is sometimes used as the basic equation to account for the effects of local topography and density variations. The difference lies in the assumption of vertical velocity distribution and consideration of density variation. The form of the governing equations changes depending on the coordinate system used, but the physical meaning of each term remains unchanged. The governing equations for NSWE in a planar Cartesian coordinate system are used here as an example.

$$\frac{\partial \eta}{\partial t} + \frac{\partial M}{\partial x} + \frac{\partial N}{\partial y} = 0 \tag{3.2}$$

$$\frac{\partial M}{\partial t} + \frac{\partial}{\partial x}\left(\frac{M^2}{D}\right) + \frac{\partial}{\partial y}\left(\frac{MN}{D}\right) = fN - gD\frac{\partial \eta}{\partial x} - \frac{D}{\rho}\frac{\partial P}{\partial x} + \frac{1}{\rho_w}\left(\tau_s^x - \tau_b^x + F_R^x\right) +$$

$$A_h\left(\frac{\partial^2 M}{\partial x^2} + \frac{\partial^2 M}{\partial y^2}\right) \tag{3.3}$$

$$\frac{\partial N}{\partial t}+\frac{\partial}{\partial x}\left(\frac{MN}{D}\right)+\frac{\partial}{\partial y}\left(\frac{N^2}{D}\right)=-fM-gD\frac{\partial \eta}{\partial y}-\frac{D}{\rho}\frac{\partial P}{\partial y}+\frac{1}{\rho_w}\left(\tau_s^y-\tau_b^y+F_R^y\right)$$

$$+A_h\left(\frac{\partial^2 N}{\partial x^2}+\frac{\partial^2 N}{\partial y^2}\right) \tag{3.4}$$

where M and N are called discharge fluxes, which are the integrated velocity from the seafloor to the water surface in the horizontal rectangular coordinates x and y, respectively. η is the water-level rise from the static water surface, D is the total depth, f is the Coriolis parameter, ρ_w is the seawater density, P is the atmospheric pressure, τ_s is the sea-surface friction force, τ_b is the bottom friction force, F_R is the radiation stress and A_h is the horizontal eddy kinematic viscosity coefficient. The radiant stress term is added to the model to incorporate the change in storm surge anomaly (approximately 10% change [e.g., Mori et al., 2009]) due to the wave setup because of wave momentum release. Wave setup is considered to have a greater effect on steeply sloping beaches facing the open ocean, where waves break rapidly. Radiation stress calculations are performed using the wave model calculations, but this term is omitted if the model is not coupled with a dynamic SSM. The sea-surface friction and bottom friction forces are expressed as follows:

$$\tau_s^x=\rho_a C_W W_x W,\ \tau_s^y=\rho_a C_W W_y W \tag{3.5}$$

$$\tau_b^x=\frac{\rho_w g n^2}{D^{\frac{7}{3}}}M\sqrt{M^2+N^2},\ \tau_b^y=\frac{\rho_w g n^2}{D^{\frac{7}{3}}}N\sqrt{M^2+N^2}, \tag{3.6}$$

where n denotes the Manning's roughness coefficient and W_x and W_y are wind speeds at 10-m height along the x- and y-directions, respectively, W is the absolute value of the wind speed at 10-m height $\left(W=\sqrt{W_x^2+W_y^2}\right)$ and C_W is the sea-surface drag coefficient.

Since the latter half of the 20th century, many studies have been conducted to estimate the sea-surface drag coefficient using wind tunnel hydraulic experiments and observations (see Table 3.2). Garratt (1977) analyzed highly reliable data obtained from various observations from 1967 to 1975 using the profile method and the eddy correlation method. They proposed an equation to calculate C_W as a function of wind speed at 10-m height. Honda and Mitsuyasu (1980) proposed a calculation formula based on experiments using a high-speed wind tunnel tank, and this formula is also cited in the storm surge inundation assumptions for Japan. Although several equations have been proposed to estimate C_W as a function of wind speed, C_W also depends on the effect of waves generated by the wind. Hence, the sea-surface drag coefficient (Janssen, 1989), which depends on wave age, is sometimes used in storm surge-wave coupled models. Powell (2006) parameterized sea-surface drag coefficients for each TC (tropical cyclone) location (right sector, left sector and rear sector) based on wind speed observation data from dropsondes.

3.2.2 Discretization

In numerical calculations, differential equations (mass and momentum conservation laws) are approximated in the form of algebraic equations with discretely defined variables, and the storm surge anomaly and flow flux at each location are calculated sequentially at certain time steps. The main discretization methods are the finite difference method (FDM), the finite volume method (FVM) and the finite element method (FEM), and the method used depends on the storm surge calculation model. An overview of these three discretization methods is provided in this section.

FDM replaces the time and spatial derivatives with function values at discrete points. The differential equation is approximated at each grid point by dividing the time and space range (i.e., computational domain) into a finite number of grid points (i.e., discrete points) and replacing the derivative term at each grid point with the function at the grid point. Therefore, an algebraic equation (difference equation) is obtained for each grid point, and the values of the variables there and at several neighboring grid points are unknown. In the case of the dynamic SSM, the values of the storm surge anomaly and the flow flux at each grid point are unknown. The application of FDM on a structural mesh is simple and effective, and is widely used. However, it can only be applied to simple geometries because it can only be used on structural meshes, which poses a significant challenge in the analysis of complex flow fields.

FVM divides the computational domain into a finite number of adjacent control volumes (CVs) and applies an integral differential equation to each CV. The storm surge anomaly and flow flux at each location are calculated sequentially by defining a grid point at the center of each CV, placing the variables to be computed there, and solving the algebraic equations for each obtained CV. FVM is adaptable to arbitrary grid geometries, allowing the generation of computational domains adapted to complex geometries and improving computational accuracy. On the other hand, FVM requires a large number of computational grids near the coastline, which increases the computational load.

The FEM divides the computational domain into a finite number of polyhedral elements and approximates each element with relatively simple algebraic equations. Specifically, the differential equations are multiplied by a weight function and integrated within the domain, and the resulting algebraic equations are solved to obtain the storm surge anomaly and discharge flux at each point. As with the FVM, this method can be applied to arbitrary geometries. However, it is difficult to find solutions for algebraic equations linearized by unstructured grids. Also, the grid generation itself is difficult.

3.2.3 Proposed dynamic SSMs

In this section, some of the previous representative dynamic SSMs are introduced. While some models are based on NSWE and perform 2D plane calculations, others perform 3D (or quasi-3D) calculations, such as the Reynolds-averaged Navier–Stokes equation, which can consider the vertical distribution of current velocity by dividing the sea surface to the seafloor into several layers. Table 3.3 summarizes basic information on each model (discretization scheme, coordinate system, input, handling of typhoons, whether a wave model is considered or not and whether tides are considered or not). For TCs, the empirical typhoon model (ETM) model uses the ETM introduced in Chapter 2 with input information on the TC track and central pressure, while the regional climate model (RCM) model uses the pressure and wind fields in the computational domain, such as the output of a meteorological model, directly input to a dynamic SSM. The RCM is based on a method that directly inputs the atmospheric pressure and wind fields in the computational domain, such as the output of a meteorological model, to a dynamic SSM.

1. SLOSH (Sea, Lake and Overland Surge from Hurricanes)
   SLOSH (Jelesnianski et al., 1992) is a dynamic SSM developed by the National Weather Service. It is the primary model that has served as the basis for other SSMs in the US. The model is mainly applied to North America (e.g., Forbes et al., 2014), and it has computed domain and bathymetry data for 37 regions (Model Basin). It is still used for real-time simulations, repeatability calculations and

probability assessments. FDM is used for discretization, which is a two-dimensional explicit method and is easy to compute. The use of a locally orthogonal polar, elliptical or parabolic coordinate system (a locally orthogonal polar, elliptical or hyperbolic coordinate system) enables it to optimize the computational domain and resolution while satisfying the CFL conditions. It can also be used to calculate storm surge overtopping and inundation of land areas. Moreover, it has many applications in risk assessment. However, the model does not consider the effects of wave setup, river discharge and flooding due to rainfall, and further improvement of the model is expected.

2. SuWAT (Surge, WAve, and Tide)

SuWAT is a NSWE-based dynamic SSM developed by Kim et al. (2008). By combining a wave model (Simulating WAves Nearshore (SWAN)) and a tidal model, it is possible to include the effects of wave setup and tidal-level fluctuations on storm surge anomalies in storm surge calculations. Parallel computation using the Message Passing Interface (MPI) can be applied to nesting to efficiently solve from the wide domain to the detailed domain. The model has been applied in many Asian countries, such as Japan and Korea, and the development team is actively updating the model. Recently, the model has been coupled with the storm surge overtopping model IFORM (Integrated Formula of Overtopping and Runup Modeling; Mase et al., 2013; Jo et al., 2024), improved to account for river flow (Toyoda et al., 2023), sewage (Jo et al., 2021) and buildings (Fukui et al., 2024).

3. ADCIRC (ADvanced CIRCulation model)

ADCIRC is a NSWE-based dynamic SSM proposed by Luettich et al. (1992). By using a triangular unstructured grid as the computational grid and applying optimized parallel computing (using MPI), ADCIRC can represent complex coastal topography, waterways and land obstacles in a computational domain of several thousand kilometer scales while reducing computational cost. FEM (spatial) and FDM (temporal) are used for discretization. The model has been widely used for inundation calculations and storm surge forecasting (e.g., Kowaleski et al., 2020) for specific coastal areas, with many applications and validations in North America, including storm surges caused by Hurricanes Katrina, Sandy and Ike (e.g., Westerink et al., 2008). The model is continuously updated by the development team, including coupling with a wave model (SWAN) (Dietrich et al., 2012) and application to global storm surge calculations (Pringle et al., 2020). The latest version is v55.

4. GeoClaw (GEOphysical Conservation LAW model)

GeoClaw (Mandli and Dawson, 2014) is a dynamic SSM developed at the University of Washington in the US. The adaptive mesh method dynamically changes the horizontal and temporal resolution by refining and merging the computational grid periodically. It differs from general nesting methods as it can optimize the computation without dividing the computational domain and does not need to consider the final domain. Therefore, this method is expected to be used in the calculation of regional storm surges caused by typhoons, considering path uncertainties. As the grid is divided and integrated using subdivision criteria, such as water-level fluctuation, flow velocity and typhoon radius, it is necessary to conduct sensitivity analysis to determine appropriate refinement criteria to balance calculation accuracy and efficiency. Additionally, the effects of artificial structures, such as levees and buildings, are not considered. The model has been applied to many cases, mainly hurricane storm surges that hit North America, and the development team is actively updating the model.

5. Delft3D

Delft3D is a 2D or 3D hydrodynamic (and transportation) model developed at Deltares. The model is characterized by the fact that it considers the vertical distribution of flow velocity using the sigma coordinate system and supports 3D calculations. The software can be applied not only to storm surge, but also to various unsteady flow and transportation phenomena, such as river currents, tsunamis, saltwater intrusion and wave-driven currents. Packages (Delft3D 4 Suite, Delft3D FM Suite 2D3D) are available for structured and unstructured meshes, in commercial (Delft3D service packages) and open source versions. The discretization differs based on the package, with the former using FDM and the latter using FVM. The commercial version comes with a license for the latest version and online support from the development team. Each package includes a module for dynamic storm surge modeling (Delft3D-FLOW and D-Flow Flexible Mesh, respectively). There are a wide range of storm surge cases from Europe, the US, Asia and other regions. The range of applications is wide, so that compound occurrence of storm surge and river flooding (Lee et al., 2023, Goulart et al., 2024) and global tide and surge reanalysis datasets (Kirezci et al., 2020) have been developed.

6. SCHISM (Semi-implicit Cross-scale Hydroscience Integrated System Model)

SCHISM is a dynamic SSM developed in Zhang and Baptista (2008). It is characterized by the use of the Reynolds-averaged Navier–Stokes equation instead of NSWE for the governing equations (i.e., no depth integration). As with Delft3D, the vertical structure of the terrain can be considered in 3D calculations. For the vertical coordinate system, the Hybrid SZ and LSC2 grids, which are based on the s and σ coordinate systems, are used, and the grid is arranged according to the topography of the seafloor. FVM or FEM is used for discretization, and an unstructured grid (triangles and quadrilaterals) is used to represent the topography, enabling seamless storm surge calculations from the ocean to the creek (creek-lake-river-estuary-shelf-ocean) to be performed efficiently. In addition to wave models, the system is coupled with sediment transportation models, water quality models and ecosystem models. SCHISM has been applied worldwide, including the US, Asia and Europe. The official website of SCHISM has a wealth of practical examples, including applications to National Oceanic Atmospheric Administration (NOAA)'s Inland-Coastal Flooding Operational Guidance System and Portugal's on-demand circulation forecasting system.

7. FVCOM (Finite Volume COmmunity Model)

FVCOM is a dynamic SSM proposed by Chen et al. (2003), Massachusetts Dartmouth University, US. Like Delft3D and SCHISM, it uses a FVM for discretization and an unstructured (triangular) mesh to represent topography. The model can accurately reflect the geometry of shorelines and breakwaters in the computational mesh. The vertical distribution of flow velocity can also be considered using the sigma coordinate system. External forces can be input as atmospheric pressure and wind fields, tidal fluctuations and river discharge. The FVCOM community has been active in developing the system, and improvements are under way. Although there are many applications worldwide, it is mostly used as an ocean model for tides, currents, marine ecosystem interactions, etc., in the US.

3.3 STOCHASTIC SSM

An example of the application of a dynamic SSM using a stochastic typhoon model is presented in this section. The results of the dynamic SSM calculations are subject to uncertainties in the model and in the input typhoon central pressure and track. Therefore, it is necessary to make probabilistic storm surge forecasts that consider these uncertainties. In this section, a method for probabilistic storm surge prediction as a stochastic SSM is defined.

P-surge (Taylor and Glahn, 2008) is a stochastic SSM used by NOAA for real-time ensemble forecasting. The model uses current and forecasted TC characteristics (i.e., typhoon intensity, size and path) published by NOAA's National Hurricane Center (NHC) official advisories to provide probabilistic information on expected storm surge while accounting for forecast errors. Specifically, probability distribution functions for (1) cross-track, (2) along-track (i.e., the storm forward translational speed), (3) TC size represented by the maximum wind speed radius and (4) TC intensity represented by the maximum wind speed are calculated using the average error over the past five years (assuming a normal distribution), and 500–1,000 representative TCs are generated. From the generated typhoons, typhoon tracks and pressure drop DelP from atmospheric pressure are extracted, and the maximum wind radius is calculated from the maximum wind speed and input into SLOSH. The probabilistic storm surge information can be computed by SLOSH by combining the likelihood of water level and typhoon path, pressure drop and maximum wind radius. The variability among the generated typhoons can be artificially increased, allowing storm surge calculations to begin without waiting for atmospheric ensemble data to be generated. There are two types of probabilistic information: the probability that the storm surge deviation exceeds a certain threshold (e.g., the probability that the storm surge deviation exceeds 2 m) and the magnitude of the storm surge deviation that exceeds a certain probability (e.g., the magnitude of the storm surge deviation with 10% exceedance probability).

A similar approach to P-surge has been applied in Japan recently, and JMA has developed the Japan Area Storm Surge Probabilistic Forecast System (PFS), which probabilistically forecasts storm surges five days in advance (JMA, 2023). The Japan Storm Surge PFS generates 21 representative TCs whose tracks are equally spaced in the lateral direction of the typhoon path, based on the typhoon forecast circle information, and performs storm surge forecast calculations up to five days ahead. The external forces are based on the atmospheric pressure and wind velocity fields from the meso-scale model owned by JMA, and the following TC bogus based on Fujita (1952) is embedded.

$$P(r) = P_\infty - \frac{P_\infty - P_c}{\sqrt{1 + \dfrac{r^2}{r_0^2}}}, \tag{3.7}$$

where r is the distance from the center, $P(r)$ is the pressure at r from the center, P_∞ is the pressure at infinity, P_c is the central pressure and r_0 is a parameter that determines the sharpness of the pressure drop. Within the high wind radius (wind speed range above 30 kt), the barometric wind field due to TC bogus is input as it is. On the other hand, outside the strong wind radius, a new pressure–wind field is calculated using TC bogus up to twice the strong wind radius and a weighted average of no wind and sea-level pressure outside the strong wind radius, and the typhoon intensity is attenuated as it moves away from the center of the typhoon. When shifting the track, TC bogus is embedded within the wind radius of each typhoon; outside the wind radius, the weighted average of the no wind (i.e., 0 m/s) and sea-level pressure and the pressure and wind speed from TC bogus are used as input. Then, to account for the uncertainty in TC translation speed, the speed of TC progression and storm surge calculations are adjusted for $21 \times 21 = 441$ cases of typhoons, and tidal effects are added to produce a storm surge probability forecast. Based on the results of this probabilistic forecast, whether a storm surge early warning should be issued is decided. While the ability to account for uncertainties in typhoon tracks through probabilistic arguments is a major advance, the problem remains that it is difficult to account for the effects of terrain friction in TC bogus, and the forecast tends to overestimate the actual values as the forecast time progresses.

The European Centre for Medium-Range Weather Forecast provides path ensemble forecasts for medium-range forecasts of TCs. Magnusson et al. (2019) performed path ensemble forecasts for Hurricanes Irma and Maria at 18-km and 5-km horizontal resolution. They succeeded in improving the accuracy of the path and intensity forecasts with the 5-km ensemble forecasts. Although the model is currently limited to probabilistic forecasting of typhoons, it is expected to be applied to stochastic SSMs in the future.

The following section outlines a stochastic SSM that evaluates the probability of storm surge based on typhoon information generated by a stochastic typhoon model. The stochastic typhoon model (e.g., Hatada and Yamaguchi, 1996; Kato et al., 2003; Hashimoto et al., 2004; Kawai et al., 2006; Yasuda et al., 2010; Nakajo et al., 2014) is a method for artificially generating typhoons using Monte Carlo simulations based on statistical characteristics, such as central pressure, translation speed and wind speed. This method has the advantage of generating typhoon datasets for thousands of years in a short period, and can obtain artificial typhoon paths and changes in central pressure that are statistically consistent with historical data. This aspect of the model has been frequently used to evaluate the impact of climate change on storm surge. As the output of the stochastic typhoon model is the track and central pressure along the track, the typhoon meteorological field represented by the ETM is used in the dynamic SSM. Therefore, as described in Chapter 2, storm surge over a wide area or over an ocean area that is greatly affected by complex topography should be calculated carefully.

3.4 DATA-DRIVEN SSM

The dynamic SSM described in Section 3.2 and the stochastic SSM for probabilistic storm surge prediction using the dynamic SSM require a large amount of computational resources because they assume a large number of typhoons and require ensemble calculations. Statistical SSMs that use regression equations based on pressure drop and wind speed have been developed to overcome these limitations. Furthermore, computational power and machine learning technology have witnessed remarkable progress, and many data-driven SSMs using artificial neural networks (ANNs) and deep learning have been proposed.

3.4.1 Statistical SSM

The improvements to the simplified storm surge estimation formula of JMA, described in Section 3.1, are discussed in this section. Yokoyama et al. (2020) used a global stochastic typhoon model (Nakajo et al., 2014) and the dynamic SSM SuWAT to improve the simplified storm surge estimation equation for the Seto Inland Sea coast. Using 1,000 years of hypothetical TC data generated by a stochastic typhoon model as input values, we performed a multitude of storm surge calculations using SuWAT and used the calculation results to improve the simplified storm surge estimation equation described in Section 3.1. To consider the effect of wave setup, the constant term (i.e., c) of Eq. (3.1) was improved by considering the effect of water-level rise owing to wave nonlinearity, as shown in Eqs. (3.7) and (3.8).

$$\zeta_{max} = a(1010 - P) + bW^2 \cos\theta + cW^2 \tag{3.8}$$

$$c = 7.890 \times 10^{-7} \times r^2 + 3.141 \times 10^{-4} \times r + 0.049 \tag{3.9}$$

where r is the distance [km] of the TC at its closest approach to the target location. The coefficients a and b are obtained by multiple regression analysis of the storm surge anomaly results from SuWAT, and c is calculated using wave height and wind speed data from JMA and Nationwide Ocean Wave information network for Ports and HarbourS (NOWPHAS) to obtain a quadratic equation for the relationship between the distance r and coefficient c at the closest approach of a TC. The obtained model, which was applied to storm surge calculations for a typhoon that hit Kobe, succeeded in improving the estimation accuracy of TCs that traveled along a path far from the target area. On the other hand, the model revealed that it is difficult to estimate the storm surge considering the time lag between the approach of a typhoon and the occurrence of a storm surge by a uniform method using the estimation formula. This is because the trend of the storm surge differs from place to place.

3.4.2 Data-driven SSMs using machine learning

From the late 2010s to 2024, there has been extensive research on storm surge prediction using machine learning. Three methods are mainly used for storm surge prediction: neural networks (ANNs), deep learning using ANNs with many intermediate layers (e.g., convolutional neural network (CNN); recurrent neural network; long short-term memory, LSTM; GAN) and regression models (e.g., ridge regression model, support vector regression, etc.). All of them are data-driven SSMs. All data-driven SSMs use typhoon time-series information (point information or spatial field) as input data, and primarily predict the time series of storm surge anomaly and the maximum storm surge anomaly. The training data can be divided into two types: those using observed tidal height data and those using calculated tidal height data from a dynamic SSM, the latter being called a surrogate model. In this section, some recent studies of data-driven

SSMs for each model are reviewed. Table 3.4 summarizes the target domain, machine learning model used, input data and training data for each data-driven SSM. For more detailed information, please refer to the respective references.

Storm surge forecasting using ANNs has existed since the early 2000s (e.g., Lee, 2006; De Oliveira et al., 2009), but owing to recent improvements in computer performance and machine learning methods, numerous research cases have been proposed from the late 2010s to 2024 and continue to be proposed. The studies of Kim et al. (2016, 2019) are introduced as research cases using observed values. We developed an ANN that predicts the time series of storm surge anomalies at a station in Tottori Prefecture using hydrological information, such as tide-level anomalies, and meteorological information, such as sea-level corrective pressure as input values at the start of the prediction. During development, the model is built based on Dreyfus in the following order: (1) find input parameters related to the predictions, (2) determine the appropriate number of hidden neurons (or units), (3) determine the best performing algorithm and associated coefficients and (4) determine appropriate functions in the hidden layer and output layer. The developed ANN was applied to storm surge forecasting at the Sakaiminato station in Tottori Prefecture, Japan, and successfully forecasted from 5, 12 and 24 h prior to the peak time (lead time), revealing that the optimal explanatory variables change with each lead time. The study of Lockwood et al. (2022) is presented as an example for applying ANN to surrogate models. Unlike Kim et al. (2016, 2019), they use only weather information (e.g., hurricane radius, wind speed, hurricane position and moving speed) as input values to construct an ANN that outputs storm surge anomalies. Using 5,018 hurricanes extracted from the National Centers for Environment Prediction (NCEP) atmospheric reanalysis data from 1980 to 2005 as input, storm surge anomalies were calculated by the dynamic SSM ADCIRC. From the obtained data, water-level time series for 300 locations in the US East and Gulf Coasts were extracted and used as training data for the ANN. The model is evaluated not only for hurricanes from NCEP atmospheric reanalysis, but also for 1,000 cases of hurricane storm surge in SSP5–8.5 downscaled by CMIP6 to confirm its high applicability for predicting climate change impacts.

Many cases of storm surge prediction using deep learning have been proposed since the late 2010s, and most of them use CNNs. Bruneau et al. (2020) and Tiggeloven et al. (2021) successfully predicted hourly storm surge anomaly time series by deep learning using the Global Extreme Sea-Level Analysis Version 2 (GESLA-2) data. The model uses meteorological information (mean sea-level pressure and wind speed at 10-m height) obtained from ERA5 atmospheric reanalysis as input, and compares ANN, LSTM, CNN and ConvLSTM (deep learning combining CNN and LSTM). Lee et al. (2021) presented an example of applying CNNs with k-mean clustering and Principal Component Analysis (PCA) to surrogate models. In this study, the North Atlantic Comprehensive Coastal Study database (NACCS database), a tropical storm surge database proposed by the US Army Corps of Engineers (Cialone et al., 2015), was used for deep learning to quickly determine the maximum storm surge anomaly. The NACCS database contains the results of calculations obtained using ADCIRC (i.e., dynamic SSM) and STeady-state spectral WAVE (STWAVE; i.e., wave model), and the authors developed a surrogate model (defined as the C1PKNet model) using the NACCS database using CNN. The explanatory variables used are the location of the low-pressure system, central pressure, speed of movement, directional angle and maximum wind radius. The developed model was validated using not only calculated storm surge values from the NACCS database, but storm surge values from three previous hurricanes (Isabel, Irene and Sandy) were also observed to confirm that the C1PKNet model predicts the maximum storm surge anomaly quickly. Important conclusions include the possibility of considering the time series of low-pressure information using CNNs, the necessity of considering bypassing storms in surrogate models and the necessity of validating past events in model evaluation.

Finally, Ayyad et al. (2022) presented a case study of forecasting using regression models, in which various regression models (ridge regression, support vector regressor, decision tree regressor, random forest and extra trees regressors and adaptive boost and gradient-boosted decision tree regressors) were used to predict the storm surge. The model outputs the maximum storm surge anomaly considering the wave setup. Four global climate models (GFDL5, HadGEM5, MPI5 and MRI5) were compared with the storm surge anomalies calculated by the dynamic SSM (ADCIRC) and wave model (SWAN) from the low pressure in the present and future climate obtained by the hurricane model proposed by Emanuel et al., which is a surrogate model to learn. The explanatory variables include the location of the typhoon (six hours before, at the time of and six hours after the arrival of the target point), maximum wind speed and distance between the eye and the target point. The reproducibility with the storm surge anomaly and return period by the dynamic SSM is good, indicating the potential of the data-driven model. However, support vector regression and AdaBoost regression are confirmed to be superior to other models as they are robust to outliers and less prone to overlearning.

3.5 COASTAL VULNERABILITY MODEL

In this section, a coastal vulnerability model for storm surge as an application example of predicting storm surge deviation as a hazard is introduced. To evaluate vulnerability, a dynamic SSM is used to calculate the inundation depth of a land area, and the obtained information, such as inundation depth, is used as an explanatory variable to calculate the damage function of an asset. Ha et al. (2021) calculated storm surge inundation and coastal vulnerability due to climate change for Osaka Bay. They conducted a cost–benefit analysis of adaptation measures for climate change (in their paper, the raising of levees). SuWAT calculates storm surge anomalies using 5,000 years of typhoon information and sea-level-rise projections generated by the global stochastic typhoon model of Nakajo et al. (2014). For land areas, inundation calculations are performed using the high-performance integrated hydrodynamic modeling system (Liang et al., 2019; Xia et al., 2019), and the resulting inundation depths are used to calculate property damage values. The obtained inundation depth is used to calculate the amount of property damage. In this study, only direct damage to property is considered, and damage amounts are calculated for three categories: residential (houses and household goods), businesses (depreciable assets and inventory assets) and public (public facilities and public services). The amount of damage for each asset is calculated using the inundation depth, exposure and damage function corresponding to the location of the asset. The relationship between inundation depth and damage rate is based on the "Cost–Benefit Analysis Guideline for Coastal Projects," which is based on the "Manual for Economic Evaluation of Flood Control Investment" by the Ministry of Land, Infrastructure, Transportation, and Tourism (Ministry of Land, Infrastructure and Transport, 2020) and the Coastal Development Effectiveness Measurement Manual in the UK. Damage calculations are performed as follows:

1. Damage to houses = total floor area of damaged building by inundation height × evaluation value per square meter by prefecture × damage rate by inundation height.

2. Damage to household items = number of affected households by inundation height × amount of household items owned per household × damage rate by inundation height.

3. Damage to depreciable assets = valuation of depreciable assets per employee of an establishment × number of employees in disaster-affected business establishments × depreciable asset damage rate by inundation height.

4. Damage to inventory assets = valuation of inventory assets per employee of an establishment × number of employees in disaster-affected business establishments × inventory asset damage rate by inundation height.

5. Damage to public = (damage to businesses + damage to residences) × 0.8225.

In this study, a cost–benefit analysis was conducted using the amount of asset damage to suggest appropriate climate change adaptation measures. The cost–benefit analysis identifies the optimal trade-off between the cost of raising levees and the avoidance of damage to assets due to storm surge inundation (benefits). The benefits are expressed as a reduction in expected annual damage cost (EADC) between the non-adaptation scenario (no embankment raising) and the adaptation scenario. EADC can be calculated for each global warming scenario Representative Concentration Pathways (RCP) by integrating the annual exceedance probability and the property damage due to storm surge inundation. The costs and benefits are expressed in Eqs. (3.10) and (3.11), respectively.

$$B_T = \sum_{t=1}^{T} \frac{B_{at}}{(1+r)^{t-1}} \tag{3.10}$$

$$C_T = C_o + \sum_{t=1}^{T} \frac{C_{at}}{(1+r)^{t-1}} \tag{3.11}$$

where B_T represents the total benefits over T years, B_{at} denotes the reduction in EADC due to the adaptation measure in year t, C_T stands for the total costs over T years, C_o represents the cost of constructing the embankment, C_{at} represents the annual maintenance costs, T represents the total investment horizon (a span of 30 years from 2030 to 2100) and r indicates the discount rate. For each adaptation measure, the net present value (NPV) and benefit–cost ratio (BCR) are calculated to assess cost-effectiveness. NPV indicates the net economic benefits over T years owing to the adaptation measure (as per Eq. 3.12). Assessing the investment effectiveness of adaptation measures can be done by comparing NPV to BCR (as per Eq. 3.13). If NPV is greater than 0, then BCR is greater than 1. This indicates that the benefits (avoidance of flood damages) from the adaptation measure outweigh the costs (embankment construction and maintenance costs), making this investment economically attractive.

$$\text{NPV} = B_T - C_T \tag{3.12}$$

$$\text{BCR} = \frac{B_T}{C_T} \tag{3.13}$$

TABLE 3.1   Values of each parameter in Eq. (3.1) and wind direction at main observation points (Hirayama et al., 2003)

Points	a	b	Wind direction	Statistic period	No. of data
Wakkanai	0.516	0.149	WNW	1960–1968	38
Abashiri	1.296	0.036	NW	1961–1968	29
Hanasaki	1.12	0.02	SE	1970–1979	38
Kushiro	1.316	0.016	SW	1954–1968	33
Hakodate	1.262	0.023	S	1957–1960	35
Hachinohe	1.429	0.015	ENE	1958–1960	7
Miyako	1.193	0.012	NNW	1945–1959	6
Ayukawa	1.346	0.02	SE	1951–1959	9
Choshi	0.622	0.056	SSW	1951–1959	6
Mera	1.935	0.012	SW	1957–1960	7
Tokyo	2.332	0.112	S 29° W	1917–1987	22
Ito	1.128	0.005	NE	1951–1966	30
Uchiura	1.439	0.024	SW	1951–1966	29
Shimizu	1.35	0.016	ENE	1951–1966	36
Omaezaki	1.324	0.024	NE	1951–1966	18
Maisaka	2.256	0.08	S	1951–1966	29
Nagoya	2.961	0.119	S 33° E	1950–1987	29
Toba	1.825	0.001	ESE	1950–1959	7
Uragami	2.284	0.025	SE	1950–1961	6
Kushimoto	1.49	0.036	S	1950–1960	10

(Continued)

TABLE 3.1 (Continued)

Points	a	b	Wind direction	Statistic period	No. of data
Shimotsu	2	0.022	SSW	1934–1960	13
Wakayama	2.608	0.003	SSW	1930–1960	12
Tannowa	2.552	0.004	SSW	1953–1960	8
Osaka	2.167	0.181	S 6.3° E	1929–1953	28
Kobe	3.37	0.087	S 24° E	1941–1987	31
Sumoto	2.281	0.026	SSE	1950–1960	10
Uno	4.109	−0.167	ESE	1950–1960	8
Kure	3.73	0.026	E	1951–1956	4
Matsuyama	4.303	−0.082	SSE	1950–1956	7
Takamatsu	3.184	0	SE	1950–1960	9
Komatsushima	1.72	0.019	SE	1951–1960	10
Kochi	2.385	0.033	SSE	1950–1960	8
Tosa-Shimizu	1.428	0.022	S	1950–1957	10
Uwajima	2.33	−0.012	SSE	1950–1956	7
Aburatsu	1.005	0.036	SE		6
Kagoshima	1.234	0.056	SSE		6
Makurazaki	0.973	0.04	S		4
Naha	1.117	0.015	N 9° E	1969–1987	19
Misumi	1.185	0.154	SSW		11
Tomie	1.094	0.027	SE		5
Shimonoseki	1.231	0.033	ESE		10
Hamada	1.17	0.021	NNW	1950–1959	6
Sakai	0.48	0.027	ENE	1950–1959	6
Miyazu	1.43	−0.014	NE	1950–1959	14

TABLE 3.2 Summary of studies on the wind drag coefficient

	Wind drag coefficient C_W	Method
Garratt (1977)	$\min\left(2\times10^{-3},\left(0.75+0.067W\times10^{-3}\right)\right)$	Observation
Honda and Mitsuyasu (1980)	$\left(1.29-0.024W\right)\times10^{-3}\left(W<8.0\left[\mathrm{m/s}\right]\right)$ $\left(0.581+0.063W\right)\times10^{-3}\left(W\geq8.0\left[\mathrm{m/s}\right]\right)$	Experiment
Large and Pond (1981)	$1.2\times10^{-3}\left(4.0\left[\mathrm{m/s}\right]<W<11.0\left[\mathrm{m/s}\right]\right)$ $\left(0.49+0.065W\right)\times10^{-3}\left(11.0\left[\mathrm{m/s}\right]\leq W\leq25.0\left[\mathrm{m/s}\right]\right)$	Observation
Wu (1982)	$\min\left(2\times10^{-3},\left(0.8+0.065W\times10^{-3}\right)\right)$	Observation

TABLE 3.3 Summary of the proposed dynamic SSMs

	SLOSH	SuWAT	ADCIRC	GeoClaw	Delft3D		SCHISM	FVCOM
					Delft3D-FLOW	D-Flow Flexible Mesh		
Discretization scheme	FDM	FDM	FEM and FDM	FVM	FDM	FVM	FVM/FEM	FVM
Coordinate system	Polar/elliptical/Hyperbolic	Cartesian	Cartesian/spherical	Cartesian	Cartesian/curvilinear/spherical	Cartesian/spherical	Cartesian/spherical	Cartesian/spherical
Grid type	Structured grid (fixed mesh)	Structured grid (fixed mesh)	Unstructured grid (fixed mesh)	Structured grid (adaptive mesh)	Structured (fixed mesh)	Unstructured grid (fixed mesh)	Unstructured grid (fixed mesh)	Unstructured grid (fixed mesh)
Model input	Topography/bathymetry meteorological information	Topography/bathymetry meteorological information	Topography/bathymetry meteorological information	Topography/bathymetry meteorological information	Topography/bathymetry meteorological information	Topography/bathymetry	Topography/bathymetry grid information Meteorological information	Topography/bathymetry grid information Meteorological information
TC implementation	ETM	ETM/RCM	ETM/RCM	ETM	RCM		RCM	RCM
Wave implementation	○ (SWAN)	○ (SWAN)	○ (SWAN)		○ (Delft3D-WAVE)	○ (D-Waves)	○ (WWIII)	○ (SWAN)
Tide implementation	○	○	○		○	○	○	○

TABLE 3.4 Summary of the proposed data-driven SSMs

	Lockwood et al. (2022)	Bruneau et al. (2020) Tiggeloven et al. (2021)	Lee et al. (2021)	Kim et al. (2019) Kim et al. (2016)	Ayyad et al. (2022)
Target region	US, US Gulf and East Coasts	Global	US Chesapeake Bay, Virginia and Maryland	Japan Sakaiminato at Tottori Coast	US New York Metropolitan Area
Used model	ANN	ANN, LSTM, CNN, ConvLSTM	CNN (with PCA and k-means clustering)	ANN	SVR, AdaBoost Regressor
Predictand variable	Surge time series (hourly, point data)	Surge time series (hourly, point data)	Peak storm surges (point data)	Surge time series (hourly, point data)	Peak storm surges considering wave setup (point data)
Predictor variable	$Lat, Lon, W,$ $V_f, R_o,$ bearing	$MSLP,$ $\Delta MSLP,$ W_x, W_y, W	$Lat, Lon, \theta,$ R_{max}, V_f, C_p	$Lat, Lon, C_p, W,$ $WD, HWS, SS,$ $SLP, DSLP$	Lat, Lon, W_{max} at ±6 or 0 hour landfall time
Model input	Time series of TC parameters (NCEP reanalysis)	Atmospheric reanalysis (ERA5 dataset)	Time series of TC parameters	Meteorological, hydrodynamic and typhoon-characteristic components (JMA dataset)	Time series of TC parameters
Training data	Numerical model results (ADCIRC @ 300 stations)	Observation (GESLA-2 @ 738 stations)	Numerical model results (NACCS-reported peak storm surge @ 3111 stations)	Observation (hydrodynamic station data at Sakai Minato)	Numerical model results (ADCIRC + SWAN)

Note: Lat: latitude of the center of TC, *Lon*: longitude of the center of TC, *W*: wind speed magnitude at 10-m height, W_x: wind speed along the *x*-direction, W_y: wind speed along the *y*-direction, W_{max}: maximum wind speed, V_f: TC translation speed, R_o: TC radius, C_p: central pressure, θ: heading direction, R_{max}: radius of maximum winds, *HWS*: highest wind speed near TC center, *SS*: storm surge height, *MSLP*: mean sea-level pressure, *SLP*: sea-level pressure, *DSLP*: drop of sea-level pressure.

REFERENCES

Ayyad, M., Hajj, M. R., & Marsooli, R. (2022). Machine learning-based assessment of storm surge in the New York metropolitan area. *Scientific Reports*, 12, 19215. https://doi.org/10.1038/s41598-022-23627-6

Bruneau, N., Polton, J., Williams, J., & Holt, J. (2020). Estimation of global coastal sea level extremes using neural networks. *Environmental Research Letters*, 15(7), 074030.

Cialone, M. A., Massey, T. C., Anderson, M. E., Grzegorzewski, A. S., Jensen, R. E., Cialone, A., … & Ratcliff, J. J. (2015). North Atlantic Coast Comprehensive Study (NACCS) coastal storm model simulations: Waves and water levels (p. 0252). Kitty Hawk, NC, USA: US Army Engineer Research and Development Center, Coastal and Hydraulics Laboratory.

Chen, C., Liu, H., & Beardsley, R. C. (2003). An unstructured grid, finite-volume, three-dimensional, primitive equations ocean model: Application to coastal ocean and estuaries. *Journal of Atmospheric and Oceanic Technology*, 20(1), 159–186. https://doi.org/10.1175/1520-0426(2003)020<0159:AUGFVT>2.0.CO;2

De Oliveira, M. M., Ebecken, N. F. F., De Oliveira, J. L. F., & de Azevedo Santos, I. (2009). Neural network model to predict a storm surge. *Journal of applied Meteorology and Climatology*, 48(1), 143–155.

Dietrich, J. C., Tanaka, S., Westerink, J. J., Dawson, C. N., Luettich, R. A., Zijlema, M., … & Westerink, H. J. (2012). Performance of the unstructured-mesh, SWAN+ ADCIRC model in computing hurricane waves and surge. *Journal of Scientific Computing*, 52, 468–497.

Forbes, C., Rhome, J., Mattocks, C., & Taylor, A. (2014). Predicting the storm surge threat of hurricane sandy with the national weather service SLOSH model. *Journal of Marine Science and Engineering*, 2(2):437–476. https://doi.org/10.3390/jmse2020437

Fujita, T. (1952). Pressure distribution within typhoon. *Geophysical Magazine*, 23(4), 437–451.

Fukui, N., Mori, N., Kim, S., Shimura, T., & Miyashita, T. (2024). Application of a subgrid-scale urban inundation model for a storm surge simulation: Case study of typhoon Haiyan. *Coastal Engineering*, 188, 104442.

Garratt, J. R. (1977). Review of drag coefficients over oceans and continents. *Monthly Weather Review*, 105(7), 915–929.

Goulart, H. M. D., Benito Lazaro, I., van Garderen, L., van der Wiel, K., Le Bars, D., Koks, E., & van den Hurk, B. (2024). Compound flood impacts from Hurricane Sandy on New York City in climate-driven storylines. *Natural Hazards and Earth System Sciences*, 24, 29–45. https://doi.org/10.5194/nhess-24-29-2024

Ha, S., Tatano, H., Mori, N. et al. (2021). Cost–benefit analysis of adaptation to storm surge due to climate change in Osaka Bay, Japan. *Climatic Change*, 169, 23. https://doi.org/10.1007/s10584-021-03282-y

Hashimoto, N., Kawai, H., Matsuura, K., & Kawaguchi, K. (2004). Development of stochastic typhoon model for performance design of coastal design. *Proceeding of the 29th International Conference on Coastal Engineering*, 4, 3615–3627.

Hatada, Y., & Yamaguchi, M. (1996). A stochastic typhoon model and its application to the estimation of extremes of storm surge and wave height. *Proceedings of the 25th International Conference on Coastal Engineering*, 2, 1389–1402.

Hirayama, H., Tsujimoto, G., Shimada, T., & Honda, N. (2003). *Coastal Engineering*, pp.60–63. CORONA PUBLISHING CO., LTD.

Honda, T., & Mitsuyasu, T. (1980). Experimental study on the effect of wind on water surface. *Proceedings of Coastal Engineering, JSCE*, 27, 90–93 (in Japanese).

Janssen, P. A. (1989). Wave-induced stress and the drag of air flow over sea waves. *Journal of Physical Oceanography*, 19(6), 745–754.

Jelesnianski, C.P., Chen, J., & Shaffer, W.A. (1992). SLOSH: Sea, Lake, and Overland Surges from Hurricanes; NOAA Technical Report NWS 48; National Oceanic and Atmospheric Administration, U.S. Department of Commerce: Silver Spring, MD, USA, pp. 1–71. Available online: https://vlab.noaa.gov/documents/6609493/36173392/SLOSH_TR48.pdf (accessed on 15 July 2024).

JMA (2023). Collection of numerical weather prediction commentaries for fiscal 2023, https://www.jma.go.jp/jma/kishou/books/nwpkaisetu/nwpkaisetu.html, accessed on March 27, 2024 (in Japanese).

Jo, J., Kim, S., Mase, H., Mori, N., & Tsujimoto, G. (2021). Development of a coupled coastal flood model of surge, wave, precipitation and sewer backflow for urban area. *Journal of Japan Society of Civil Engineers, Ser. B2 (Coastal Engineering)*, 77(2), I_253–I_258 (in Japanese).

Jo, J., Kim, S., Mori, N., & Mase, H. (2024). Combined storm surge and wave overtopping inundation based on fully coupled storm surge-wave-tide model. *Coastal Engineering*, 189, 104448.

Kato, F., Torii, K., & Shibaki, H. (2003). Extreme statistics analysis of wave overtopping rate by a stochastic typhoon model. *Proceedings of Coastal Structures 2003 Conference*, 520–527. https://doi.org/10.1061/40733(147)43

Kawai, H., Hashimoto, N., & Matsuura, K. (2006). Improvement of stochastic typhoon model for the purpose of simulating typhoons and storm surges under global warming. *Proceeding of the 30th International Conference on Coastal Engineering*, 2, 1838–1850.

Kim, S., Matsumi, Y., Pan, S., & Mase, H. (2016). A real-time forecast model using artificial neural network for after-runner storm surges on the Tottori coast, Japan. *Ocean Engineering*, 122, 44–53.

Kim, S., Pan, S., & Mase, H. (2019). Artificial neural network-based storm surge forecast model: Practical application to Sakai Minato, Japan. *Applied Ocean Research*, 91, 101871.

Kim, S. Y., Yasuda, T., & Mase, H. (2008). Numerical analysis of effects of tidal variations on storm surges and waves. *Applied Ocean Research*, 30(4), 311–322.

Kirezci, E., Young, I.R., Ranasinghe, R. et al. (2020). Projections of global-scale extreme sea levels and resulting episodic coastal flooding over the 21st Century. *Scientific Reports*, 10, 11629. https://doi.org/10.1038/s41598-020-67736-6

Kowaleski, A. M., Morss, R. E., Ahijevych, D., & Fossell, K. R. (2020). Using a WRF-ADCIRC ensemble and track clustering to investigate storm surge hazards and inundation scenarios associated with Hurricane Irma. *Weather and Forecasting*, 35(4), 1289–1315. https://doi.org/10.1175/WAF-D-19-0169.1

Large, W. G., and S. Pond (1981). Open Ocean Momentum Flux Measurements in Moderate to Strong Winds. J. Phys. Oceanogr., 11, 324–336, https://doi.org/10.1175/1520-0485(1981)011<0324:OOMFMI>2.0.CO;2.

Lee, T. L. (2006). Neural network prediction of a storm surge. *Ocean Engineering*, 33, 483–494.

Lee, J. W., Irish, J. L., Bensi, M. T., & Marcy, D. C. (2021). Rapid prediction of peak storm surge from tropical cyclone track time series using machine learning. *Coastal Engineering*, 170, 104024.

Lee, W., Sun, A. Y., Scanlon, B. R., & Dawson, C. (2023). Hindcasting compound pluvial, fluvial and coastal flooding during Hurricane Harvey (2017) using Delft3D-FM. *Natural Hazards*, 1–30.

Liang, Q., Ming, X., & Xia, X. (2019). A high-performance integrated hydro-dynamic modelling system for real-time flood forecasting. *38th IAHR World Congr - "Water Connect World"*, 38, 5246–5255. https://doi.org/10.3850/38wc092019-0969

Lockwood, J. W., Lin, N., Oppenheimer, M., & Lai, C.-Y. (2022). Using neural networks to predict hurricane storm surge and to assess the sensitivity of surge to storm characteristics. *Journal of Geophysical Research: Atmospheres*, 127, e2022JD037617. https://doi.org/10.1029/2022JD037617

Luettich, R. A., Westerink, J. J., & Scheffner, N. W. (1992). ADCIRC: An advanced three-dimensional circulation model for shelves, coasts and estuaries, Report 1: Theory and methodology of ADCIRC-2DDI and ADCIRC-3DL. Tech. Rep. DRP-92-6, U.S. Army Corps of Engineers, 137 pp. [Available from ERDC Vicksburg (WES), U.S. Army Engineer Waterways Experiment Station (WES), ATTN: ERDC-ITL-K, 3909 Halls Ferry Rd., Vicksburg, MS 39180-6199.]

Magnusson, L., Bidlot, J. R., Bonavita, M., Brown, A. R., Browne, P. A., De Chiara, G., … & Malardel, S. (2019). ECMWF activities for improved hurricane forecasts. *Bulletin of the American Meteorological Society*, 100(3), 445–458.

Mandli, K. T., & Dawson, C. N. (2014). Adaptive mesh refinement for storm surge. *Ocean Modelling*, 75, 36–50.

Mase, H., Tamada, T., Yasuda, T., Hedges, T. S., & Reis, M. T. (2013). Wave runup and overtopping at seawalls built on land and in very shallow water. *Journal of Waterway, Port, Coastal, and Ocean Engineering*, 139(5), 346–357.

Ministry of Land, Infrastructure and Transport (2020) Manual of Economic Evaluation of Flood Control Investment (In Japanese).

Mori, N., Takada, R., Yasuda, T., Mase, H., & Kim, S. Y. (2009). Effects of vertical mixing due to strong wind condition on storm surge. *Journal of Japan Society of Civil Engineers, Ser. B2 (Coastal Engineering)*, 65(1), 241–245 (in Japanese).

Nakajo, S., Mori, N., Yasuda, T., & Mase, H. (2014). Global stochastic tropical cyclone model based on principal component analysis and cluster analysis. *Journal of Applied Meteorology and Climatology*, 53(6), 1547–1577.

Powell, M. D. (2006). Drag coefficient distribution and wind speed dependence in tropical cyclones. Final Report to the National Oceanic and Atmospheric Administration (NOAA) Joint Hurricane Testbed (JHT) Program.

Pringle, W. J., Wirasaet, D., Roberts, K. J., & Westerink, J. J. (2020). Global storm tide modeling with ADCIRC v55: Unstructured mesh design and performance. (No Title).

Taylor, A., & Glahn, B. (2008). Probabilistic guidance for hurricane storm surge. Proc. 88th AMS Annual Meeting, New Orleans, Louisiana, USA, 20–24 January 2008, https://ams.confex.com/ams/88Annual/webprogram/Paper132793.html (last access: 16 May 2023), 2008.

Tiggeloven, T., Couasnon, A., van Straaten, C., Muis, S., & Ward, P. J. (2021). Exploring deep learning capabilities for surge predictions in coastal areas. *Scientific Reports*, 11(1), 17224.

Toyoda, M., Mori, N., Kim, S., Shibutani, Y., & Yoshino, J. (2023). Assessment of compound occurrence of storm surge and river flood in Ise and Mikawa Bays, Japan using a framework of atmosphere–ocean–river coupling. *Natural Hazards*, 120, 3891–3917.

Westerink, J. J., Luettich, R. A., Feyen, J. C., Atkinson, J. H., Dawson, C., Roberts, H. J., Powell, M. D., Dunion, J. P., Kubatko, E. J., & Pourtaheri, H. (2008). A basin- to channel-scale unstructured grid hurricane storm surge model applied to Southern Louisiana. *Monthly Weather Review*, 136(3), 833–864. https://doi.org/10.1175/2007MWR1946.1

Wu, J. (1982). Wind-stress coefficients over sea surface from breeze to hurricane, J. Geophys. Res., 87(C12), 9704–9706, doi:10.1029/JC087iC12p09704.

Xia, X., Liang, Q., & Ming, X. (2019). A full-scale fluvial flood modelling framework based on a high-performance integrated hydrodynamic modelling system (HiPIMS). Advances Water Resources, 132, 103392. https://doi.org/10.1016/j.advwatres.2019.103392

Yasuda, T., Mase, H., & Mori, N. (2010). Projection of future typhoons landing on Japan based on a stochastic typhoon model utilizing AGCM projections. *Hydrological Research Letters*, 4, 65–69.

Yokoyama, K., Yasuda, T., Kim, S. Y., Nakajo, S., & Shimura, T. (2020). Study on statistical prediction method of storm surges in the Seto Inland Sea employing stochastic typhoon model. *Journal of Japan Society of Civil Engineers, Ser. B2 (Coastal Engineering)*, 76(2), 1087–1092 (in Japanese).

Zhang, Y., & Baptista, A. M. (2008). SELFE: A semi-implicit Eulerian-Lagrangian finite element model for cross-scale ocean circulation. *Ocean Modelling*, 21(3-4), 71–96.

CHAPTER 4

Features and limitations of the ETM and its latest applications

4.1 SIMPLE ENSEMBLE EXPERIMENTS FOR STORM SURGE (SEES) FORECASTING USING AN EMPIRICAL TYPHOON MODEL (ETM)

In this section, the characteristics and limitations of storm surge forecasting based on ETMs is described. The paper by Toyoda et al. (2021) is used as an example for detailed discussion. From this section, you will be able to understand the characteristics of an ETM and understand the conditions in which it can be applied.

First, an overview of the study and the computational methods employed are presented. The paper discusses the limitations of storm surge forecasting based on a combination of empirical models and storm surge models (SSMs) for Typhoon Haishen in 2020, which originated in the Northwest Pacific and developed to a minimum central pressure of 920 hPa and maximum wind speeds of 50 m/s. Haishen, which developed in the East China Sea, developed into a powerful typhoon that maintained its strength while reaching a maximum wind speed of 50 m/s. It was followed by a strong storm in the Northwest Pacific. The typhoon moved northward over the East China Sea and approached the Kyushu area in Japan while maintaining its strength. This study introduces storm surge forecasting using the Weather Research and Forecasting (WRF) regional climate models (RCM)

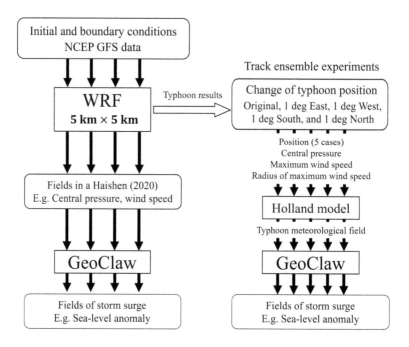

FIGURE 4.1 Computational flow of SEES. (Toyoda et al. 2021)

and Holland's parametric typhoon model. First, the results of the SSM (GeoClaw) using the WRF climate field and the results of the parametric model are compared (Figures 4.1 and 4.2).

Then, forecast experiments were conducted under four conditions with different starting times: four days before (4 DB), three days before (3 DB), two days before (2 DB) and one day before (1 DB). The reference date and time was 12:00 UTC on 6 September 2020, which is the time when Typhoon Haishen was closest to the Kyushu region.

Next, this study compares the results of the forecast simulations using the Holland model (Parametric Tropical Cyclone model (PTC), Figure 4.3; dashed lines), which is used in practice, with those using the coupled model (Figure 4.3; solid lines). The forecast values by the Japan Meteorological Agency (JMA) (3 DB, 2 DB and 1 DB) were used for the input condition of PTC location and intensity. The radius of maximum wind speed (R_w) value was set to be 70 km based on the satellite estimation value by National Oceanic and Atmospheric Administration (NOAA). The results of both methods revealed that the accuracy of the storm surge scale generally tends to improve as the forecast period reduces. Comparing the results for the same date showed that the coupled simulation has a higher

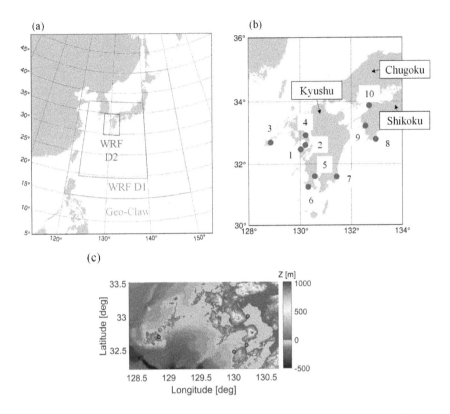

FIGURE 4.2   Simulation fields of WRF and GeoClaw. (a) The green line indicates the GeoClaw simulation area, the red line indicates the WRF domain 1 simulation area and the blue line indicates the WRF domain 2 simulation area. (b) The blue points indicate storm surge observation sites. 1: Reihoku, 2: Kuchinotsu, 3: Fukue, 4: Oura, 5: Kagoshima, 6: Makurazaki, 7: Aburatsu, 8: Tosa-Shimizu, 9: Uwajima, 10: Matsuyama. (c) This is topography data with 270-m resolution of GeoClaw. The red points correspond to No.1–4 in (b). (Toyoda et al., 2021)

accuracy in the storm surge scale. PTC overestimates the storm surge by approximately 0.30 m (30%, even with a one-day advance forecast), while the coupled model forecasts a storm surge of a reasonable value based on the observed value. This is because PTC does not consider the attenuation of the typhoon, which shows the limitation of this method. On the other hand, the peak time forecast tends to be different from the scale forecast, and the PTC is closer to the 3 DB and 2 DB forecasts. This is owing to the error caused by the delay in the approach time of the typhoon in the initial and boundary values input to the WRF in the coupled model. Therefore, it is necessary to prepare a large number of tracks to account for uncertainty

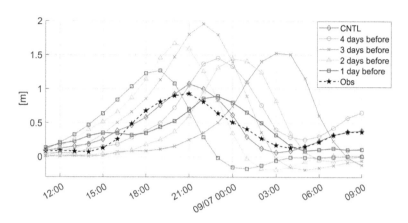

FIGURE 4.3   Time series of storm surges at the Port of Oura. The vertical axis represents the height of the storm surge, and the horizontal axis represents time. The color solid lines indicate the results of WRF–GeoClaw coupled model, and the color dashed lines indicate the results of PTC model. (Toyoda et al., 2021)

in the forecast error. In this case, the typhoon track was a straight line from south to north, and PTC was able to calculate it with high accuracy (especially the peak time of storm surge). However, the error is likely to be larger when the track is more complicated, such as when it is a curved line.

Even with an empirical model forecast, the time of peak storm surge can be obtained by preparing a large number of path patterns. However, this report tends to overestimate the magnitude of storm surges, indicating the limitations of the method. The typhoons in the northwestern Pacific Ocean tend to move westward in the tropics, then turn around owing to westerly winds from the mid latitudes, and finally move northeastward. The empirical model is able to predict typhoons with linear paths well, but the error may increase when the trajectory is complex.

This chapter also describes a simple path ensemble storm surge forecast that utilizes WRF weather forecasts as a compromise. To evaluate the impact of different tracks on storm surge forecasts, the results of simple ensemble experiments for storm surge forecasting (SEES) assuming multiple typhoon tracks based on dynamic tropical cyclone (TC) forecasts are discussed using the parametric Holland typhoon model to increase the number of TC tracks. Moreover, the characteristics and limits of storm surge forecasts were determined. First, the simulation method of SEES is described. In this study, the Holland (1980) model, a parametric TC model for pressure and wind, was used to set up the meteorological field.

The parametric typhoon model assumes an axis-symmetric pressure field given a minimum central pressure and R_m, and it has the advantage of arbitrary typhoon track translation and a very low computational cost. Therefore, the TC track ensemble can be easily considered using the parametric Holland typhoon model. This method is widely used by many municipalities in Japan for storm surge inundation assumption. However, it has a significant disadvantage: the dynamical and local characteristics (e.g., pressure at a specific point) of the typhoon and the intensity and temporal changes in R_w cannot be considered. In this study, the temporal changes in the typhoon intensity and R_w were considered using the hourly simulation result of the Global Forecast Grids (GFS) obtained with the WRF model as the typhoon information for input into the parametric Holland typhoon model (Figure 4.1). Atmospheric pressure and wind speed, according to Holland (1980), were used in GeoClaw (Mandli and Dawson, 2014).

The parametric Holland typhoon model requires information on the typhoon position (latitude and longitude), central pressure, maximum wind speed and R_m. In this study, the results of forecast experiments (4 DB, 3 DB, 2 DB and 1 DB) performed with the WRF model were used as the input conditions for the parametric Holland typhoon model. Furthermore, the output of the WRF model was used to obtain the values of the central pressure, maximum wind speed and R_m (every hour). SEES combines highly accurate meteorological model values with a low computational cost parametric model. Therefore, it is a realistic method in terms of accuracy and computational cost.

Regarding SEES, the perturbation of the track was set to 1°. Five types of typhoon track perturbations were considered: original track, 1° E, 1° W, 1° S and 1° N. Therefore, storm surge forecasts were obtained for each day (4 DB, 3 DB, 2 DB and 1 DB) using the five typhoon tracks and the same intensity and R_m (total 20 forecasts). The perturbation was chosen as 1° because the averaged error of JMA's 24-h prior forecast of the typhoon track was estimated to be approximately 100 km. It should be noted that the typhoon track was shifted by 1° from start to end.

Storm surge forecasting using the empirical model was conducted using the abovementioned approach. WRF–GeoClaw coupled calculations and parametric model simulations were performed to simulate storm surges caused by typhoons. Although the coupled simulation has a high computational cost, it can represent a realistic storm surge. However, PTC, which has poor accuracy, tends to overestimate the storm surge. In this section, we propose a method that can be used by practitioners to reduce computational cost while ensuring accuracy by applying the WRF output to the parametric model. We evaluated the practicality and limitations of this method using SEES.

First, as discussed in the previous section, we confirmed the time series of the storm surge height in Oura (Figure 4.4). In any simulation of 4 DB–1 DB, the peak time is the earliest in the case of 1° N. The case of 1° N shows the largest maximum storm surge height in all three cases except in the 1 DB forecast. The initial position shift of the typhoon to the north caused an approach to western Kyushu before the typhoon intensity weakened. The largest peak storm surge height in the 1 DB forecast was in the case of 1° E. The typhoon track shifted to the west by approximately 30 km in the WRF simulation results for the 1 DB. Hence, the shift of the track to the east caused the Ariake Sea to be located near the R_m of the typhoon. Therefore, a stronger wind flow to the Ariake Sea led to a significant increase in storm

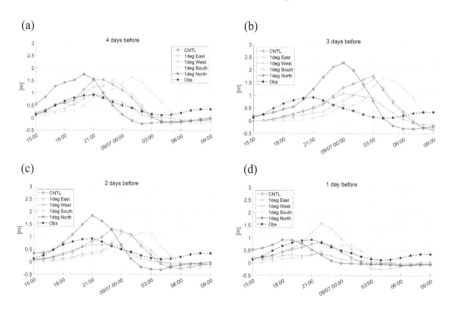

FIGURE 4.4 Time series of storm surges at the Port of Oura by SEES ((a) 4 DB, (b) 3 DB, (c) 2 DB and (d) 1 DB). The vertical axis represents the height of the storm surge, and the horizontal axis represents time. The red diamond line is original from the WRF results, the orange circle line represents moving the typhoon track east at 1°, the blue X line represents moving the typhoon track west at 1°, the green triangle line represents moving the typhoon track south at 1° and the purple square line represents moving the typhoon track north at 1°. (Toyoda et al. 2021)

surge height. The peak times and maximum storm surge fluctuated in the range of 5 h and 1.0 m, respectively, for the five typhoon tracks in Oura.

Storm surges were forecasted in western and eastern Kyushu, and parts of the Chugoku–Shikoku region, which is far away from the typhoons. There were ten tidal gauges affected by Typhoon Haishen in the Kyushu and Chugoku–Shikoku regions (Figure 4.2). The maximum storm surge heights and peak occurrence times were compared between the SEES and gauge data, as shown in Figure 4.5. Note that the observed values in this discussion are the average values for the 3-min observations published by JMA, which are different from the 1-hour observed values used in the time series analysis. The vertical axis in Figure 4.5 represents the maximum storm surge height at each site. The X, star and horizontal axes represent the observed value, ensemble mean of SEES and peak time error relative to the observed value, respectively. Note that, for Tosa-Shimizu and Uwajima, the maximum values were recorded more than 1.5 days (2160 min) before the observed peak in some cases. However, the errors were too large to be represented on the same axis. Therefore, Figure 4.5 does not show errors of more than 720 min. Comparing the observed values with the ensemble mean (Figure 4.5) shows that the peak intensity and time of the storm surge can be forecast with high accuracy in Reihoku and Kuchinotsu because X and star are close to each other (within 20% error). In Fukue and Oura, the storm surge height errors are 0.16 m and 0.37 m, respectively. However, the peak time errors shown in Figure 4.5 are forecasted as 221 and 303 min, respectively, which are significant delays. Therefore, the tendency of maximum storm surge forecast errors is small at Reihoku and Kuchinotsu, which are close to the typhoon center, and large in other sites (especially Aburatsu, Matsuyama, Uwajima and Tosa-Shimizu, which are farther from the typhoon). At the every simulation site, at least one member showed that can forecast result with the small error for the peak time of the storm surge. However, no ensemble member was successful in accurately forecasting the maximum storm surge height at a site far from the typhoon center.

Next, we focus on the variations in the peak time error and maximum storm surge errors at each site (Figure 4.6). Figure 4.6 shows the absolute average error of the WRF–GeoClaw coupled simulation and SEES results. The averages were weighted according to the forecast time, and the weight was increased for the 1 DB forecast. In Reihoku, which has the best forecast accuracy, the average time error is 59 min and the average storm surge error is 0.18 m. On the other hand, in Tosa-Shimizu, which

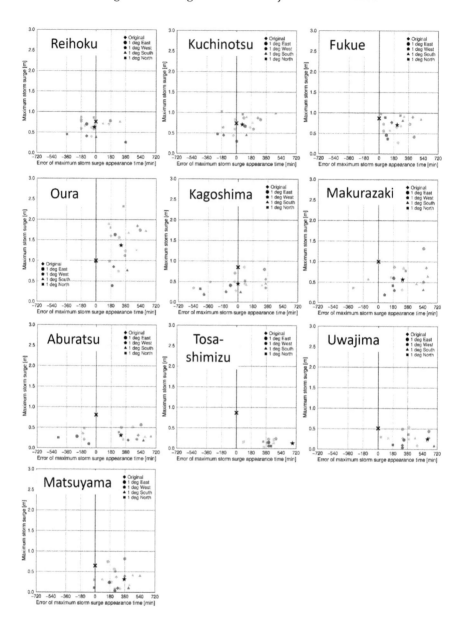

FIGURE 4.5 Scatter plot of maximum storm surge occurrence time error and storm surge deviation at each site. Orange represents the result of 4 DB, blue represents 3 DB, green represents 2 DB and purple represents 1 DB. The shape of the symbol is the same as that in Figure 9. X represents the observed value, and the star represents the average value of all cases. Note that the results are not displayed in the figure for simulations with a time error of 720 min or more. (Toyoda et al., 2021)

has the worst accuracy, the average peak time error is 694 min and the storm surge error is 0.75 m. The coupled model and SEES tend to have smaller errors at sites close to the typhoon and larger errors at sites that are far from the typhoon. The large errors in the storm surge scales for Makurazaki and Tosa-Shimizu are owing to the relatively large storm surges that occurred and smaller simulated values. Although the mean error of SEES is larger than that of the coupled model, the trend of the error is similar between the coupled model and SEES, indicating that the accuracy of SEES storm surge forecasting can be ensured even though it is based on a parametric model. The accuracy of the storm surge scale by SEES was improved compared to using only the parametric typhoon model. As shown in Figure 4.5, all the sites contain at least 1 case out of the 20 cases that exhibit a peak time close to the observed peak time, and it is possible to forecast the peak time regardless of the distance from the center of the typhoon to the target site. In other words, the peak of the storm surge can be forecast regardless of the distance from the typhoon in several cases. However, in the maximum storm surge forecast, almost all the cases were underestimated by 40% or more, except for four cites (Reihoku, Kuchinotsu, Fukue and Oura). Therefore, a successful forecast (e.g., within 10% error) of the maximum storm surge was not obtained. Thus, the performance of SEES is superior to that of WRF–GeoClaw in forecasting the peak time of the storm surge than in forecasting the maximum storm surge height. Furthermore, the usefulness of SEES for translation of the track is effective at sites near the typhoon (western Kyushu), and only the peak time forecast is effective at sites far from the typhoon. Additionally, even at sites close to the typhoon, the variability of each case tends to be greater in the inner bays (e.g., Oura) than in the bays facing the open ocean. Moreover, the impact of typhoon tracks on storm surges is greater in inner bays than in bays facing the open ocean. In this study, we used the results of four WRF simulations to perform 20-track ensemble forecasts of storm surges. If a greater number of WRF simulations can be performed, the SEES method may be able to accurately forecast the scale of storm surges at sites far from the typhoon.

This calculation confirms that the ensemble of 20 patterns improves the probability of capturing the peak time of the storm surge. Nevertheless, the accuracy is still poor with respect to storm surge magnitude and storm surge forecasting in areas far from the typhoon. To resolve such errors, a dynamic model that can consider the off-axis nature of typhoons that make a landfall in Japan and the effect of topography is required.

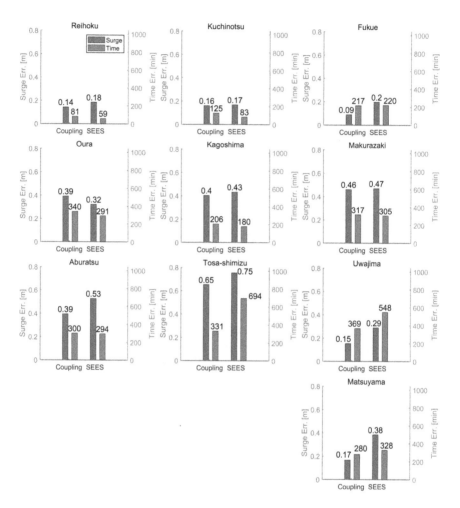

FIGURE 4.6 The averages of the error by WRF–GeoClaw coupled model and SEES for the ten sites of the storm surge calculation. Blue and red indicate the maximum storm surge error and peak time errors, respectively, with the left and right vertical axes corresponding to storm surge error and time error. (Toyoda et al., 2021)

4.2 OPTIMIZATION OF ETM PARAMETER SETTING

As described in Chapter 2, Holland's PTC has a scaling parameter B, which should be determined by the value of R_m. In Japan, this parameter is often assumed to be 1, but it is important to determine it according to the shape of the typhoon. The authors previously examined the relationship between the scaling parameter B and past typhoons to develop an equation to estimate the scaling parameter based on the relationship between typhoon intensity and distance between the typhoon and the port. This analysis shows that the radius ratio, the ratio of the maximum pressure gradient radius Rp to the maximum wind speed radius R_m, which determines the scaling parameter, is related to typhoon attenuation.

This is a discussion about the characteristics of R_p/R_w for the simulated 49 typhoon landfall cases in Japan from 2000 to 2017, as shown in Figure 4.7. We focused on the peak times and landfall times of 49 typhoons. Moreover, the same analysis was conducted on 20 ECs that occurred in 2020 to compare with the 49 typhoons. The National Center for Environmental Prediction (NCEP) Final analysis (FNL) data (1 deg×1 deg) were used to analyze the structure of the EC. In this study, ECs corresponding to the following three conditions were defined as the target EC. (1) Extratropical cyclones do not originate from typhoons. This analysis focuses on cases that have been generated and developed exclusively as ECs. (2) Extratropical cyclones that developed to less than 980 hPa were selected as targets. (3) There are no other disturbances (typhoons or ECs less than 980 hPa) around the target. Extratropical cyclones satisfying these conditions are detected in the same area (120°–180° E, 10°–70° N), similar to the typhoon case. According to the frequency distribution, many typhoon cases show $R_p/R_w=0.9$ at the peak time of the typhoon (blue bars in Figure 4.8). This result shows that R_p and R_w are almost the same at the peak time of typhoons (red bars in Figure 4.8). The number of cases of R_p/R_w tends to be larger than 0.4 and less than 0.6 at the time of landfall (red). Approximately half of the typhoon landfall cases featured $\sim R_p/R_w=0.5$. Hence, the fixed value of R_p/R_w (0.5) in the ETM is not an irrelevant setting. Furthermore, it cannot be ignored because each class of 0.6–1.0 has the ratio of approximately 10%. However, the R_p/R_w of the typhoon at the time of landfall varies significantly from case to case. To eliminate data bias, the 20 ECs were extracted with different values of minimum pressure, season and location of occurrence. Consequently, the results vary. However, the largest class has 0.1–0.2 frequency distribution, and 75% of the total is included in the classes less than 0.6. Therefore, R_p/R_w is smaller in ECs than in typhoons. Focusing on the shape of the frequency distribution, the peak of the frequency of R_p/R_w at the peak time of the typhoon is located on the far right. Furthermore, the peaks of the frequency distribution move to the left in the order of the landfall times of typhoons and ECs. Therefore, it is inferred that the value of R_p/R_w becomes smaller than the typhoon peak times, landfall times. A TC undergoing extratropical transition (ET) is denoted here as ET-TC. The widely used constant relationship

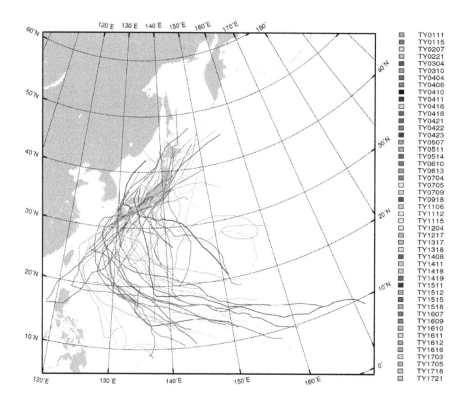

FIGURE 4.7   Observed tracks of target typhoons from genesis to decay, according to the JMA best track.

of the R_p/R_w by the ETM is not satisfied in the simulated TCs and can be changed from offshore to the coast owing to the developing process of TC and topographical effects on TCs. The frequency distributions show a tendency for R_p/R_w to decrease in the order of peak intensity of typhoons, landfalls and ECs. During landfall, R_p/R_w is smaller than that of the peak because the typhoon structure is closer to that of an EC.

As discussed above, the radius ratio is believed to be linked to intensity decay and the collapse of the typhoon structure, and the empirical model setup used in storm surge forecasting should incorporate this change. Although the radius ratio is not publicly available information, the two estimating equations presented in this study are presented together. There could be a negative correlation between the relative weakening rate (Wrate [%]) of the typhoon intensity after the peak time and R_p/R_w. Additionally, the intensity of the longer-traveled typhoon tends to decay from its peak. Hence, there could also be a negative correlation between

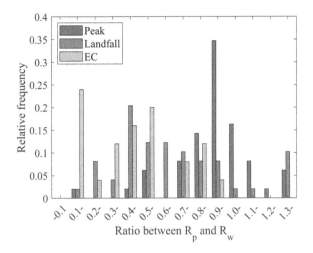

FIGURE 4.8 Distribution of ratio between R_p and R_w in 49 simulated typhoons (the blue bar indicates R_p/R_w at the time of peak, the red bar indicates R_p/R_w at the time of landfall and the yellow bar indicates R_p/R_w of extratropical cyclones). (Toyoda et al., 2022a)

the distance from the point where the typhoon reaches peak intensity to landfall (D_{pl} [km]) and R_p/R_w. Note that D_{pl} is a straight-line distance between two points, not the distance along the track of the typhoon.

The optimization of parameter B as a function of R_p/R_w is expected to improve the accuracy of the ETM. In the previous section, the relationship of R_p/R_w was analyzed, and R_p/R_w was found to be related to W_{rate} and D_{pl}. To evaluate the R_p/R_w of a typhoon expected to landfall in the future, the predicted W_{rate} and D_{pl} are required. In this study, we propose an empirical relationship that uses the typhoon forecast values (typhoon intensity and typhoon track) provided by the JMA to predict W_{rate} and D_{pl}. Two different empirical relationships are proposed to estimate R_p/R_w as a function of W_{rate} or D_{pl}:

$$R_p/R_w = -0.6743 \times \tanh(0.2884 \times W_{rate}) + 1.0, \tag{4.1}$$

$$R_p/R_w = -\tanh(0.4639 \times D_{pl}/10^3 + 0.11) + 1.11. \tag{4.2}$$

Here, these equations are used as a robust estimation method that reduces the weight of outliers by the least-squares method because R_p/R_w varies significantly. Additionally, tanh is determined as the functional form considering the shape of the scatter plot, and there is no typhoon case wherein

R_p/R_w is 0. The coefficients and intercept are calculated under the condition that R_p/R_w is 1 when W_{rate} or D_{pl} is 0. The application of this equation tends to improve the estimation errors for the pressure and wind speed distributions at landfall.

The new parameter B is obtained from Eqs. (2.21), (4.1) and (4.2). Here, the new parameter B is used to validate the improvement in ETM. The effect of improvement using the new parameter B was validated based on three typhoon cases that severely damaged Japan recently (Figure 4.9). The typhoon cases used for validation were Typhoon Jebi (2018), Typhoon Faxai (2019) and Typhoon Hagibis (2019). Moreover, each validation was conducted based on time series at three sites ((a) Kansai International Airport, (b) Chiba and (c) Yokohama), where significant pressure drop and strong wind speed were observed for each typhoon (Figure 4.9; black line). The simulations were conducted using the time information when the typhoon was the closest in each case. Simultaneously, the estimation results using the conventional ETM ($B=1$, B calculated using Eq. (2.8) by Holland) are also shown for comparison (Figure 4.9; red and yellow lines). The best track of the JMA and the estimated value of R_w by satellite observation of NOAA (https://rammb-data.cira.colostate.edu/tc_realtime/) are used for the values of typhoon position, typhoon intensity and R_w required for calculation using ETM.

As shown in Figure 4.9, the accuracies were improved in some cases. The blue dashed line is the result obtained using Eq. (4.1), and the green dashed line is the result obtained using Eq. (4.2). At the Kansai International Airport at the time of the Jebi attack (Figure 4.9a), the parameter B was estimated to be 1.12 using Eq. (2.8), 0.850 using Eq. (4.1) and 0.755 using Eq. (4.2). This accurately represents the pressure until the typhoon approaches. The average estimation error of air pressure is approximately 7.0 hPa and 8.4 hPa in the setting of $B=1$ and $B=1.12$, respectively. However, the average error of the ETM with the new parameter B is 4.9 hPa and 3.5 hPa, respectively. Furthermore, even if the new parameter B is used, the time lag of the wind speed cannot be improved. The waveform is considerably different from the observed value. The shape of Typhoon Jebi had already collapsed into an ellipse when Jebi reached Kansai International Airport. Therefore, the wind speed of Jebi cannot be expressed by the ETM. This problem shows the limitation of the ETM that assumes that the typhoon structure is approximated to be axisymmetric. Therefore, it is necessary to operate a dynamic model, such as a mesoscale meteorological model, to solve this problem.

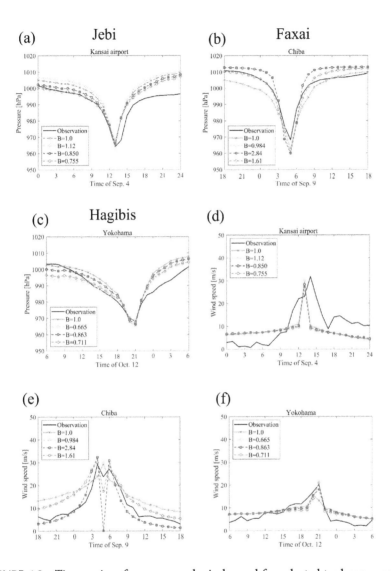

FIGURE 4.9 Time series of pressure and wind speed for selected typhoon events. Panels (a)–(c) show the pressure and panels (d)–(f) show wind speeds. Panels (a) and (d) are results of the Kansai International Airport for the case of Typhoon Jebi. (b) and (e) are results of the observation site at Chiba for the case of Typhoon Faxai. (c) and (f) are the results of the observation site at Yokohama for the case of Typhoon Hagibis (the black solid line indicates observation by JMA. The red dashed line with x marks indicates the estimation with B = 1. The yellow dashed line with diamonds indicates the estimation with the parameter B using the estimation Eq. (2.8). The blue dashed line with square indicates the estimation with the new parameter B using the estimation Eq. (4.1). The green dashed line with circles indicates the estimation with the new parameter B using the estimation Eq. (4.2). (Toyoda et al. 2022a)

In the case of Faxai (Figure 4.9b), the parameter B was estimated to be 0.984 using Eq. (2.8), 2.84 using Eq. (4.1) and 1.61 using Eq. (4.2). There was no significant improvement (0.24 hPa) in the pressure time series. However, the accuracy improved significantly in the time series of the wind speed. Two peaks that are not expressed by the conventional model ($B = 1$ and $B = 0.984$) can be expressed, and it can be said that the effect of parameter B is large. The average estimation errors of wind speed were approximately 4.8 m/s and 4.9 m/s in the setting of $B = 1$ and $B = 0.984$, respectively. However, the average errors in the ETM with the new parameter B were −2.1 m/s and 3.5 m/s, respectively. B should be set larger than 1 because Faxai made a landfall immediately after reaching its peak intensity.

Finally, in the case of Hagibis (Figure 4.9c), the parameter B is estimated as 0.665 using Eq. (2.8), 0.863 using Eq. (4.1) and 0.711 using Eq. (4.2). The accuracy of air pressure and wind speed improved. Particularly, the peak of the wind speed improved significantly. B estimated using Eq. (4.1) tends to be larger than that estimated using Eq. (4.2). A larger B indicates that the typhoon is closer to its peak intensity, and, thus, the pressure gradient is sharper. The three typhoon cases have different intensities and tracks. B estimated using Eq. (2.8) tends to be more accurate than the method that uses $B = 1$ permanently. Furthermore, B estimated using Eqs. (4.1) and (4.2) are more accurate than that estimated using Eq. (2.8), considering the changes in pressure and wind speed when the typhoons approach. The accuracy of typhoon intensity improved by employing the novel method that determines parameter B using the constructed estimation Eqs. (4.1–4.2) and Eq. (2.20) proposed by Holland. However, it should be noted that the statistical equations developed in this study are only applicable to typhoon landfalls exclusively on the Japanese Islands. Although there are still limitations to the application of the empirical model, improvements are under way.

4.3 EMPIRICAL PSEUDO-GLOBAL WARMING EXPERIMENT (E-PGWE) METHOD USING ETM

As mentioned at the beginning of this chapter, the intensity of TCs is expected to increase owing to the recent global warming. Therefore, it is necessary for port administrators and practical fields to adapt to such changes in the environment and take measures for disaster prevention that may occur in the future. In the future, it is desirable that all calculations, including those for typhoon weather and storm surge, be performed using dynamic models. However, this is currently impractical owing to the high computational cost and expertise required. Therefore, the authors devised a method to empirically estimate typhoon intensity under future climate conditions and calculate future changes in storm surge using information that is easily available to anyone (e-PGWE).

e-PGWE is based on the future change trends obtained from statistical analysis of the results of pseudo-warming experiments for typhoons that made a landfall in Japan between 2000 and 2017. The future change trends for a large number of typhoons have revealed that two physical quantities, "time from onset to landfall (Tl)" and "maximum wind speed radius at landfall (R_{ml})," have a significant impact on typhoon intensity in the present climate. Two types of equations are used to estimate future changes: typhoon intensity and maximum wind speed radius.

The future changes in the intensity and structure of landfalling typhoons will be significantly affected by T_l and R_{ml} at landfall times in the Japanese Islands, and there are significant differences between low T_l cases (small R_{ml} cases) and high T_l cases (large R_{ml} cases). Therefore, the following estimated equation was obtained using a nonlinear regression analysis with future change in P_c (ΔP_c) as the objective variable and T_l and R_{ml} as explanatory variables, based on the high-resolution typhoon database. The estimation formula is based on the discussion in Section 3.3, and the best combination is cubic for T_l and linear for R_{ml}. We constructed the formula by nonlinear regression analysis using the results obtained in Section 3.3 and the generalized additive model. The magnitude of estimation errors (bias, root mean square), correlation coefficient, degree of freedom-adjusted coefficient of determination (CD) and Akaike's information criterion (AIC) are used to determine whether the estimated equation is the best combination of T_l and R_{ml}. The estimation formula (Eq. 4.3) has the highest CORR (0.64) and CD (0.42) and lowest AIC among all the considered combinations.

$$\Delta P_c = -0.0304T_l^3 + 0.759T_l^2 - 3.73T_l - 0.243R_{ml} + 8.60 \qquad (4.3)$$

This formula (regression coefficients) was constructed using robust estimation to remove outliers or noise in the data and bound the calibration signal within narrow limits. However, a constant indicates residuals. This equation consists of an additive model of a cubic equation for T_l and a linear equation for R_{ml}. T_l and R_{ml} are parameters under the present climate. T_l obtained from the JMA best track data and R_{ml} estimated using Eq. (2.1) are input into Eq. (4.3). P_c in the future climate is calculated as the sum of the observed P_c in the present climate and future change ΔP_c obtained using Eq. (4.3). Figure 4.10 provides a comparison of the estimated P_c for empirical-pseudo-global warming downscaling (e-PGWDs) performed using Eq. (4.3) with d-PGWDs performed using High-resolution typhoon model (HTM). The obtained CORR is as high as 0.71. Thus, it is possible to capture the overall trend with a simple method. However, there are bias errors in describing typhoons weaker than 980 hPa as stronger and typhoons stronger than 940 hPa as weaker. Furthermore, there must exist non-negligible uncertainty with a root mean square error (RMSE) of 17.3 hPa. Overall, although it is necessary to consider the uncertainty of the estimation formula, the formula expressed in Eq. (4.3) is expected to estimate P_c under the future climate with a certain degree of accuracy. The standard error in Eq. (4.3) is 9.87 hPa. The CORR between T_l and R_{ml} is −0.1, which suggests that multicollinearity in the equation is not a problem.

In addition to typhoon intensity P_c, typhoon structure R_{ml} may change under the future climate conditions because of global warming. Therefore,

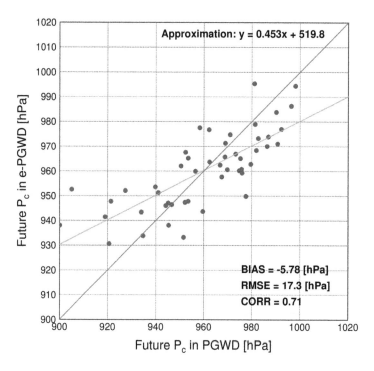

FIGURE 4.10 Scatter diagram of future-climate central pressure (P_c) in PGWDs (horizontal axis) and e-PGWDs (vertical axis). (Toyoda et al., 2022b)

it is necessary to consider future changes in P_c and R_{ml} in the storm surge estimation for the future climate. As R_{ml} is a required input condition for ETM, the following estimation formula was obtained using a nonlinear regression analysis with the future change in the typhoon structure ΔR_{ml} as the objective variable, and T_l and R_{ml} as explanatory variables, in the same process to Eq. (4.3) for ΔP_c.

$$\Delta R_{ml} = -0.0263 T_l^3 + 0.338 T_l^2 + 2.40 T_l - 0.648 R_{ml} + 31.2 \qquad (4.4)$$

where ΔR_{ml} is the future change in the radius of the maximum wind speed (km) and T_l and R_{ml} are taken as explanatory variables under the present climate. Moreover, constant means residuals. For ΔP_c, the regression equation is constructed by performing a robust estimation. This equation also consists of an additive model of a cubic equation for T_l and a linear equation for R_{ml}. The relationship between ΔR_{ml} and T_l (Figure 4.11 a) and that between ΔR_{ml} and R_{ml} under the present climate (Figure 4.11b) show that the data are fitted using cubic and linear functions, respectively. The obtained CORR is 0.49 and RMSE is 25.4 km. Although the correlation of the R_{ml} formula is inferior to that of the P_c formula, it is considered that the tendency of future changes is captured. Eq. (4.4) has a standard error of 23.1 km.

In this study, we estimate the typhoon meteorological field under the future climate using $P_c + \Delta P_c$ and $R_{ml} + \Delta R_{ml}$ as input parameters for the

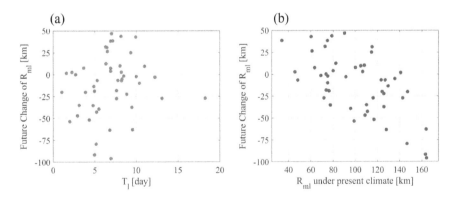

FIGURE 4.11   (a) The scatter plot between T_l and future change of R_{ml} (ΔR_{ml}). The scatter plot between R_{ml} under present climate and future change of R_{ml} (ΔR_{ml}). The correlation coefficient is 0.24 for T_l and ΔR_{ml} and −0.61 for R_{ml} under the present climate and ΔR_{ml}. (Toyoda et al., 2022b)

TABLE 4.1   Attribute variables for three target typhoons in 2018 (Typhoons Jongdari, Jebi and Trami) used for e-PGWD: input typhoon parameters (P_c, T_l and R_{ml}), estimated future typhoon changes (ΔP_c and ΔR_{ml}) and estimated future typhoon parameters (P_{cf} and R_{mlf}) with standard errors of estimation (P_{ce} and R_{mle})

		Typhoon Jongdari (2018)	Typhoon Jebi (2018)	Typhoon Trami (2018)
Target Bay (Target point)		Ise Bay (Port of Nagoya)	Osaka Bay (Port of Osaka)	Ise Bay (Port of Nagoya)
Input	P_c [hPa]	970	950	960
	T_l [day]	4.17	7.13	9.21
	R_{ml} [km]	113.9	77.9	109.3
Future change	ΔP_c [hPa]	−23.6	−9.36	−11.7
	ΔR_{ml} [km]	−28.6	+5.47	−9.40
Estimated values	P_{cf} [hPa]	946.4	940.6	948.3
	$\pm P_{ce}$	±9.87		
	R_{mlf}	85.3	83.4	99.9
	$\pm R_{mle}$	±23.1		

ETM, as evaluated by Eqs. (4.3) and (4.4). To account for the uncertainties in Eqs. (4.3) and (4.4), in addition to the case where the estimates of ΔP_c and ΔR_{ml} are used directly, we compute the cases of adding and subtracting the standard errors ($P_{ce} = 9.87$ hPa and $R_{mle} = 23.1$ km) ($P_c + \Delta P_c$ and $R_{ml} + \Delta R_{ml}$). (The abovementioned results are presented as $\pm P_{ce}$ and $\pm R_{mle}$ in the tables). In other words, we perform computations for a total of nine cases combining three patterns of P_{ce} (−9.87, 0.0 and +9.87 hPa) and three patterns of R_{mle} (−23.1, 0.0 and +23.1 km). Thus, we conduct the future-climate experiments for storm surges using ETM and SSM, while considering the uncertainties of the e-PGWD method.

We considered the three cases of Typhoons Jongdari (2018), Jebi (2018) and Trami (2018) to verify the e-PGWD validity and calculated their future changes (Table 4.1). The T_l and R_{ml} values for those typhoons are obtained from the JMA best track as input parameters described in Table 4.1. Then, future changes ΔP_c and ΔR_{ml} are estimated using Eqs. (4.3) and (4.4) for each case. Typhoon Jongdari with small T_l and large R_{ml} tends to have a large negative future change in P_c. In the case of Typhoon Jebi, where T_l is large and R_{ml} is small, the future change in typhoon intensity is not considerably large. When calculating ETM, the standard errors of the estimation formula for ΔP_c and ΔR_{ml} are also added and subtracted to consider uncertainties for the future change. Overall, nine calculations for each typhoon cases were performed considering the combination of three patterns of

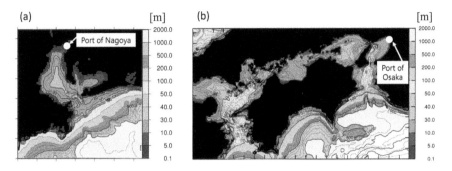

FIGURE 4.12 Horizontal distributions of the ocean topography around (a) Ise Bay and Mikawa Bay, and around (b) Osaka Bay are used in the SSM. The red points indicate the target ports for validating and discussing the SSM results: (a) Port of Nagoya and (b) Port of Osaka. (Toyoda et al., 2022b).

ΔP_{ce} (−9.87, 0.0 and +9.87 hPa) and three patterns of ΔR_{mle} (−23.1, 0.0 and +23.1 km). For the storm surge calculation, the d-PGWDs and e-PGWDs obtained from the SSM are evaluated and validated at the Port of Nagoya for Typhoons Jongdari and Trami and the Port of Osaka for Typhoon Jebi (the red points in Figures 4.12a and b). The total calculation period was 24 h from before to after the landfalls of the typhoons. Our ensemble calculations for the validation of e-PGWD assume that the typhoon track is not changed from the present climate, and nine types of input values are prepared for the typhoon intensity and typhoon radius. Moreover, nine types of storm surge results can be obtained in each typhoon case.

The results will be discussed hereafter. First, we confirm the reproducibility of the present-climate experiments for Typhoons Jongdari, Jebi and Trami (Figure 4.13). The maximum sea-level anomalies reproduced by HTM are 0.75, 2.85 and 1.33 m for Typhoons Jongdari, Jebi and Trami, respectively. The maximum sea-level anomalies observed by JMA are 0.73, 2.77 and 1.44 m, respectively, which indicate that the comparison between the HTM and observations accurately reproduces the sea-level anomalies that occurred. Next, the results obtained using the e-PGWD method are validated by comparing them with the results obtained using the d-PGWD approach. The maximum sea-level anomalies given by d-PGWD are 1.22, 3.08 and 2.05 m for Typhoons Jongdari, Jebi and Trami, respectively. Considering the uncertainties in the ΔP_c and ΔR_{ml} estimates obtained by e-PGWD, the ensemble means (standard deviations) of the maximum sea-level anomalies were found to be 1.74 m (0.52 m), 3.26 m (0.74 m) and 1.98 m (0.54 m) for Typhoons Jongdari, Jebi and Trami, respectively.

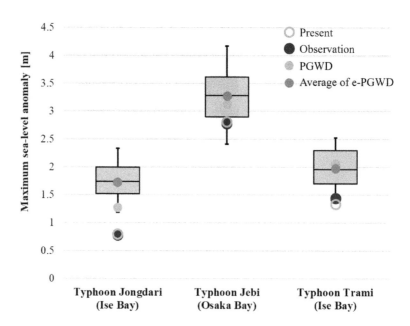

FIGURE 4.13 Box-and-whisker plot of the maximum sea-level anomalies at the target port simulated by e-PGWDs for nine cases. The blue, dark-blue, orange and red points indicate the maximum sea-level anomalies at the target port simulated by the observations (black points), present-climate experiments (green points), PGWD results by HTM (orange points) and average values for all e-PGWDs. The box height includes half of all the ensemble sizes in e-PGWDs, and the black line in the box represents the median value. (Toyoda et al., 2022b).

The maximum sea-level anomalies estimated using d-PGWD were distributed within the standard deviation of using e-PGWD in all cases (the whiskers in the boxplot). Therefore, the e-PGWD method based on the constructed estimation formula yields reasonable results. Thus, although this verification is insufficient because there were only three cases, the possibility to estimate the future changes of storm surges of different scales using e-PGWD was confirmed.

Finally, based on the results of previous studies, we describe the characteristics and limitations of storm surge forecasting using empirical models. A simple path ensemble storm surge prediction method that can be applied to practical situations using the empirical model is also presented. The results show that the empirical model has the advantages of simple and fast storm surge prediction and the ability to predict the time of peak occurrence if the variation of the path is considered. On the other hand, the empirical model has large errors in predicting the scale of the storm surge and areas far from the typhoon, which is a limitation of the current model. Next, the modification of the scaling parameter is introduced as a room for improvement of the empirical model. The scaling parameter was set based on the maximum wind speed radius and maximum pressure gradient radius of the typhoon, and in most cases, 1 is used. However, it is expected that the accuracy of local pressure and wind speed will be improved by setting values that are tailored to individual typhoon features by calculating backward based on the intensity decay and distance from the port. Finally, an empirical pseudo-warming experiment method that can be implemented in an empirical model is also presented. This is a simple method to estimate the future change in a typhoon in the future climate (the end of the 21st century) by obtaining the time from onset to landfall and the maximum wind radius in the present climate and inputting them into the estimation equation. Storm surge calculations can be performed using the results of these estimations without needing to run a specialized meteorological model.

REFERENCES

Holland, G. J. (1980). An analytic model of the wind and pressure profiles in hurricanes. *Monthly Weather Review*, 108(8), 1212–1218. https://doi.org/10.1175/1520-0493(1980)108<1212:AAMOTW>2.0.CO;2

Mandli, K. T. and C. N. Dawson. 2014. "Adaptive mesh refinement for storm surge." *Ocean Modelling*, 75, 36–50. doi:10.1016/j.ocemod.2014.01.002.

Toyoda, M., Fukui, N., Miyashita, T., Shimura, T., & Mori, N. (2021). Uncertainty of storm surge forecast using integrated atmospheric and storm surge model: A case study on Typhoon Haishen 2020. *Coastal Engineering Journal*. https://doi.org/10.1080/21664250.2021.1997506

Toyoda, M., Mori, N., & Yoshino, J. (2022a). Optimization of empirical typhoon model considering the difference of radius between pressure gradient and wind speed distributions. *Coastal Engineering Journal*. https://doi.org/10.1080/21664250.2022.2035514

Toyoda, M., Yoshino, J., & Kobayashi, T. (2022b). Future changes in typhoons and storm surges along the Pacific coast in Japan: Proposal of an empirical pseudo-global-warming downscaling. *Coastal Engineering Journal*, 64(1), 190–215. https://doi.org/10.1080/21664250.2021.2002060

CHAPTER 5

Future focus of ETM and storm surge forecasting and conclusion

In this book, empirical typhoon models (ETMs) for typhoon forecasting, computational methods for storm surge forecasting and case studies of storm surge forecasting using ETM have been described in detail. This chapter concludes with a discussion of the prospects for disaster forecasting using ETM, considering the recent changes in the subject of research.

5.1 NEW APPROACH TO ETM

ETM is expected to continue to be used as a practical method in many regions, including Japan, because ETM is a reasonable way of reproducing peak tropical cyclones (TCs) by its nature (see Chapter 2 for details). In the Atlantic region, Holland's ETM is expected to continue to operate. In the Atlantic region, the Holland model has been continuously updated and can reproduce meteorological fields with high accuracy by using settings optimized for the region. In the Northwest Pacific, some previous studies have applied the generic algorithm (GA) to the typhoon weather field to obtain optimal typhoon parameters for the environmental field and to run numerical models (Lee et al., 2006; You et al., 2012; Gong et al., 2022). These studies have reported that GAs can be applied to meteorological models (Lee et al., 2006), storm surge models (SSMs) (You et al., 2012) and empirical models (Gong et al., 2022) with high efficiency to derive

 DOI: 10.1201/9781003478782-5

model parameters that are appropriate for the environmental field. Gong et al. (2022) reported that the model parameters applied to the ETM can be efficiently derived for the typhoon meteorological field. This study uses GAs when estimating R_w, an ETM parameter of Jelesnianski (1965), which is also introduced in Chapter 2 of this book.

> GA starts with a population that represents a potential solution to a problem. A population consists of a specified number of individuals encoded by genes. Each individual is actually a characteristic chromosome entity. As the main carrier of genetic material, chromosomes are the collection of multiple genes. The initial population is generated randomly. Then, the generation evolves successively producing better approximate solutions. Each generation selects individuals based on their fitness in the problem domain. Genetic operators are applied to select, crossover, and simulate mutation processes in natural genetics to produce populations representing new solution sets.

R_w is obtained by setting the range of α and β in Eq. (2.15) and iterating until the values are constant. The error is evaluated by comparing the fit with the observed data during the typhoon until the optimal combination of parameters α and β are obtained.

An effort to calculate rainfall for TCs based on statistical trends (R-CLIPER) is being conducted in parallel (Marks et al., 2002; Tuleya et al., 2007). In the R-CLIPER model, a climatological rainfall rate is determined and then integrated along the storm track. Because the primary interest in TC rainfall is over land, the variation in rainfall rate after landfall needs to be considered. Additionally, a number of studies have shown that the rainfall rate is a function of storm intensity, with a tendency for higher rain rates for stronger storms. This effect will also be considered in the R-CLIPER model. The starting point for R-CLIPER development is the hourly rain gauge data from the primary and secondary stations in the US, available from the National Climatic Data Center archive. Data from the cooperative observer network are also available in this archive for approximately 2,500 sites within the US and its territories. These data were obtained for nearly all US landfalling tropical storms and hurricanes from 1948 to 2000. This sample includes 120 storms, of which 63 were hurricanes and 57 were tropical storms just prior to landfall.

R-CLIPER is based on the exponential relationship between rain rate and radius; the gauge rain rate as a function of distance from the storm center r and time t subsequent to landfall—GRR(r, t)—was modeled with

$$GRR(r,t) = A(t)R < r_m \tag{5.1}$$

and

$$GRR(r,t) = A(t)\exp\left[-\frac{r - r_m}{r_e}\right] r \geq r_m, \tag{5.2}$$

where $A(t)$ is a function describing the temporal function of the rain rate (see below), r_m is the radial extent of the inner-core rain rate and r_e is a measure of the radial extent of the tropical system rainfall.

Based on hurricane behavior, the temporal variation of the rain rate $A(t)$ was modeled out to be 500 km as

$$A(t) = ae^{-\alpha t} + b, \tag{5.3}$$

where a and b are empirically determined constants. The functional form of GRR in Eqs. (5.1)–(5.3) has five free parameters (r_m, r_e a, b and α). These parameters were obtained from least-squares fit to the binned rain rate data as a function of r and t, where r ranged from 0 to 500 km and t ranged from 0 to 48 h.

Statistical models have been developed not only for pressure and wind fields, but also for precipitation, and considering flooding caused by rainfall and storm surge caused by strong wind during typhoons will be possible in the future. On the other hand, as introduced in Chapter 2, simpler methods utilizing artificial intelligence (AI) are also being developed. In addition to existing ETM, AI-based estimation and correction methods are expected to be actively developed.

5.2 NEW APPROACH TO STORM SURGE FORECASTING

As for SSMs, although there are no fundamental differences, such as changes in the governing equation system, there are active efforts to improve the efficiency of calculations from conventional structured mesh to unstructured mesh and adaptive mesh. Unstructured mesh and adaptive mesh methods are expected to be applicable to cross-regional cases because they allow for a wider computational domain. Generally, the development of computers and improvement in observation techniques is expected to enable faster and higher-resolution forecasting in the future.

Furthermore, faster and more accurate forecasting by stochastic SSM has been proposed recently. In P-surge, as introduced in Chapter 3, several representative values of TC intensity and location are defined and then combined in all possible ways to generate TCs needed for the storm surge ensemble forecast known as factorial sampling (Taylor and Glahn, 2008). Kyprioti et al. (2021) used quasi-Monte Carlo simulations with low-discrepancy sequences (Lemieux, 2009) for numerical integrals related to probabilistic integrals to improve the computational efficiency with the same accuracy as that of factorial sampling. Jung et al. (2023) proposed a method in which a Monte Carlo simulation is performed to generate the TCs needed for the ensemble each time the National Hurricane Center (NHC) updates its advisory on TCs. Here, to improve the computational efficiency of the Monte Carlo simulation, adaptive importance sampling across storm advisories is used with the advisory provided by the NHC at the time of computation. As discussed above, a faster and more accurate stochastic SSM has been constructed, and further development of real-time storm surge prediction is expected.

Additionally, several methods utilizing sub-grid models that can incorporate the effects of microtopography which cannot be resolved in a computational grid, artificial structures such as buildings and other factors with low cost and high accuracy have been proposed for studies up to inundation calculations. For example, a research group at the University of Notre Dame (Kennedy et al., 2019; Begmohammadi et al., 2021, 2023) has added terms related to integral properties of the fine-scale topography to the standard long wave equation to represent complex bathymetry, coastal topography and water channel. Moreover, they proposed a system of shallow water equations with sub-grid information. The system needs to be closed by variables defined on a coarse grid, such as averaged fluid velocity, so that they are related to the added terms for sub-grid information. The developed model has been applied to storm surge calculations in various complex terrains and has succeeded in improving the calculation accuracy over the standard long wave equation, even with a coarse grid. Fukui et al. (2022a, 2024) proposed a sub-grid model for urban storm surge inundation simulations, in which a group of buildings is collectively represented by a coarser grid than the buildings (e.g., grid size is the order of 10 or 100 m). In the sub-grid model, the drag force per unit area acting on a group of buildings is calculated based on the number of buildings in the grid, average projected area in east–west and north–south directions and average building height. Then, the calculated drag forces are fed back to the velocity field via momentum conservation. The number of buildings in the grid, the average projected area and the average building height are defined as SGS parameters (sub-grid-scale parameters), which are calculated from the 3D shape data of buildings and high-resolution topographic data, and are characterized by the fact that they consider SGS building information on a coarser grid. The developed sub-grid model has been validated in the case of an idealized numerical experiment using simple urban topography and historical storm surge inundation events (TC Haiyan in 2013). It was shown that the sub-grid model can represent inundation characteristics, such as inundation depth and inundated area, consistent with building-resolving simulation using high-resolution grid and field survey results. In conventional research methods, models that solve for inundation at higher resolution are common as computer performance improves, but the time required for a single calculation and the high cost of the equipment are problematic. On the other hand, it has been reported that a sub-grid model can represent inundation with the same accuracy as a high-cost model, even with a low-cost calculation at a reduced resolution. In the same way that ETMs are used for meteorological fields, simple and efficient models have been proposed for inundation calculations. Such technology is necessary to eliminate bottlenecks for practitioners to perform inundation calculations in the future.

The recent improvement in survey technology has also contributed to the development of high-resolution topographic data, and Light Detection And Ranging (LiDAR) and laser scanning have enabled us to obtain elevation data, including microtopography (e.g., in Pradhan and Kim, 2015; Zhao et al., 2015), which can be applied to storm surge calculations to improve accuracy. Moreover, building data is also being enhanced. Japan's Ministry of Land, Infrastructure, Transportation, and Tourism has launched Project PLATEAU to develop 3D city models that reproduce the urban space of each city. They have released 3D city models for 56 cities as of 2020. The 3D city models are based on the international standard CityGML, which stores attributes, such as building shape, use and material according to the level of detail (LOD). LODs are assumed to range from LOD1 to LOD4, with LOD1 storing building shape and height information, LOD2 storing roof shape in addition to LOD1 information, LOD3 storing windows and doors (openings) in addition to LOD2 information and LOD4 storing interior information in addition to LOD3 information. Currently, in many cases, LOD2 and up are already in place. Such detailed building information has the potential to be applied to storm surge inundation prediction. For example, inundation calculations that directly reflect the topography of buildings have been studied (e.g., Blumberg et al., 2015; Takagi et al., 2016), and the sub-grid model by Fukui et al. (2022a) mentioned above was used to calculate sub-grid information. A 3D city model by PLATEAU (MLIT, 2021) has been used (Fukui et al., 2022b). In terms of visualization of storm surge forecast results, applications of Cross Reality (XR) technology, such as virtual reality (Fujimi and Fujimura, 2020; Simpson et al., 2022), have emerged in recent years, and it is expected that technology to reproduce storm surge in a real urban space, including 3D city models, will also be developed. Thus, technologies to reproduce storm surges in real urban spaces, including 3D city models, will be developed further.

Data-driven approaches are being actively developed for highly efficient and accurate storm surge forecast/prediction, as mentioned in Chapter 3. Particularly, a huge number of synthetic TC data generated by stochastic typhoon models enable to develop more accurate and robust surrogate models. Then, they are expected to be applied in practice.

The authors' opinion about the prospective of the storm surge forecast using ETM is stated. As mentioned in Chapter 1, typhoon disasters are expected to become more severe in the future owing to the effects of global warming. Particularly, more intensive typhoons cause not only wind storms but also heavy rainfall. In such cases, it is necessary to forecast storm surge as in the past, and it is also required to be able to calculate precipitation using ETM in a simple manner. Although ETMs that extend the scope of application to precipitation are still in the development stage, it is expected to calculate precipitation using ETM for general typhoon precipitation in specific areas by making full use of statistical information. The ability to calculate precipitation is expected to expand the use of ETM for typhoon surge disasters and typhoon flood disasters and for predicting and validating compound disasters.

5.3 CONCLUSION OF THIS BOOK

This book focuses on ETMs, which are widely used in practice. Additionally, various research cases on ETMs and the characteristics and limitations of storm surge prediction using ETMs are discussed in detail. The contents of this book are summarized in the following chapters.

5.3.1 Chapter 1

Chapter 1 describes the target phenomenon of this book, TCs. In particular, the characteristics of TCs and the extensive damage caused by them in recent years are discussed. TCs are intensifying in various oceanic regions, such as the Atlantic, the Northwest Pacific and the Indian Oceans, causing disasters every year even though the level of disaster prevention has improved. TCs are closely related to global warming, which has been progressing in recent years. Studies on the increase in intensity due to the rise in sea-surface temperature, decrease in the number of TCs and change in their migration speeds are introduced. Similar studies are being conducted as national projects, and it is expected that more robust future predictions will be obtained in the future.

5.3.2 Chapter 2

In Chapter 2, we introduced several representative types of ETMs and their concepts, which are the main theme of this book. Among them, the pressure distribution equation by Schloemer (1954) can be regarded as the prototype of many ETMs proposed today. Many ETMs have been proposed since then, including those by Jelesnianski (1965), Holland (1980, 2010) and Fujii and Mitsuta (1986), and their accuracy has improved. Recently, AI has been used to set the optimal ETM parameters. AI is also used to calculate optimal settings. AI has also been used to directly estimate meteorological fields, and is likely to be used more widely in the future.

5.3.3 Chapter 3

In Chapter 3, the authors presented an overview of the past findings, frontlines of research and practice about SSMs. Additionally, various SSMs are introduced in detail. Statistical models based on tide observation data used to be utilized in practice, but numerical models based on dynamic models, such as nonlinear shallow water equations (NSWEs), are the main methods nowadays because of the improvement in computing technology. Originally, the finite difference method (FDM) discretization scheme with fixed structure mesh was mainstream. Additionally, the adaptive mesh refinement method and unstructured grid method to reduce the computational cost and to represent complex coastal topography, waterways and land obstacles have been proposed. Moreover, the coupling with other models, such as wave and tide models, has been conducted. From the late 2000s, the stochastic approach has been practically used as dynamic SSM calculations are subject to uncertainties in the dynamic model and the characteristics of TC. Therefore, stochastic SSMs, such as P-surge, have been developed and utilized to provide probabilistic storm surge forecasts that consider these uncertainties. From the late 2010s to 2024, the advancement in machine learning technology, particularly deep learning, has been enhanced, thereby developing data-driven SSMs. Particularly, artificial neural network (ANN)- or convolutional neural network (CNN)-based surrogate models trained with a large number of numerical results obtained from dynamic SSMs are proposed. Thus, storm surge forecasts considering probabilistic information and data-driven approach will be used more widely in the future. The coastal vulnerability model is also important to evaluate damage by storm surges. Ha et al. (2021) calculated storm surge inundation and coastal vulnerability owing to climate change for Osaka Bay. They conducted a cost–benefit analysis for adaptation measures to climate change. Thus, the coastal vulnerability models have the potential to be used for coastal hazard risk and for providing and assessing adaptation strategies.

5.3.4 Chapter 4

In Chapter 4, the authors present a case study in which the ETM and SSM described in the previous chapters are actually used in a case analysis. First, a simple track ensemble storm surge forecast for Typhoon Haishen, which hit Japan in 2020, is described by considering several tracks provided using the ETM. In this case study, the output of the 3D weather model WRF (Weather Research and Forecasting model) was used as the input field to the ETM, and the ever-changing values of atmospheric pressure and maximum wind speed radius were considered. Although the ETM method showed a certain level of accuracy for the storm surge scale near the typhoon and the time of the storm surge peak, it had difficulty in predicting accurately at locations far from the typhoon and the storm surge scale. This indicates the limitation of ETMs.

As an example of ETM parameter optimization, Schloemer (1954), Holland (1980) and Fujii and Mitsuta (1986) introduced a study on the value of scaling parameter B, which is also used in ETMs. This parameter is fixed as $B = 1$ in Japan's storm surge inundation assumptions, and many studies have highlighted its problem. In fact, the estimation of B as a function of the distance between the center of the typhoon and the port and as a function of the typhoon attenuation rate has improved the accuracy of the meteorological fields. Similar optimization studies for other parameters, not limited to parameter *B*, are expected to enable highly efficient and accurate reproduction of the meteorological field.

The last part of Chapter 4 introduces an effort to conduct an empirical pseudo-global warming experiment (e-PGWE) using an ETM based on the statistical trends of a large number of typhoon cases. This effort is based on the results of PGWE for about 18 years of typhoons that have hit Japan. Statistical equations are proposed based on the trend that "the number of days from genesis to landfall" and "the radius of maximum wind speed at landfall in the present climate" have a significant impact on the future changes (intensity and radius) of typhoons. Furthermore, the future changes obtained by the equations were used to estimate the future typhoon meteorological field in the ETM. Using this method, it is possible to project the future storm surge without requiring specialized knowledge or high-cost simulations.

The above contents are a detailed description of ETM and storm surge forecasting. We hope that the information presented in this book will assist readers in their ETM operations. The research results on ETM and storm surge forecasting are constantly being updated, and there are many useful case studies that are beyond the scope of this book. Note that the ETMs and SSMs presented in this book and the case analyses using them are only some of the research examples.

REFERENCES

Begmohammadi, A., Wirasaet, D., Silver, Z., Bolster, D., Kennedy, A. B., & Dietrich, J. C. (2021). Subgrid surface connectivity for storm surge modeling. *Advances in Water Resources*, 153, 103939.

Begmohammadi, A., Wirasaet, D., Poisson, A., Woodruff, J. L., Dietrich, J. C., Bolster, D., & Kennedy, A. B. (2023). Numerical extensions to incorporate subgrid corrections in an established storm surge model. *Coastal Engineering Journal*, 65(2), 175–197.

Blumberg, A. F., Georgas, N., Yin, L., Herrington, T. O., & Orton, P. M. (2015). Street-scale modeling of storm surge inundation along the New Jersey Hudson River waterfront. *Journal of Atmospheric and Oceanic Technology*, 32(8), 1486–1497.

Fujii, T., & Mitsuta, Y. (1986). Simulation of winds in typhoons by a stochastic model. *Journal of Wind Engineering*, 28, 1–12.

Fujimi, T., & Fujimura, K. (2020). Testing public interventions for flash flood evacuation through environmental and social cues: The merit of virtual reality experiments. *International Journal of Disaster Risk Reduction*, 50, 101690.

Fukui, N., Mori, N., Miyashita, T., Shimura, T., & Goda, K. (2022a). Subgrid-scale modeling of tsunami inundation in coastal urban areas. *Coastal Engineering*, 177, 104175.

Fukui, N., Mori, N., Kim, S. Y., Shimura, T., & Miyashita, T. (2022b). Efficient numerical modeling of storm surge inundation over metropolis using individual drag force model and adaptive mesh refinement. *Journal of Japan Society of Civil Engineers, Ser. B2 (Coastal Engineering)*, 78(2), I_229–I_234 (in Japanese).

Fukui, N., Mori, N., Kim, S., Shimura, T., & Miyashita, T. (2024). Application of a subgrid-scale urban inundation model for a storm surge simulation: Case study of typhoon Haiyan. *Coastal Engineering*, 188, 104442.

Gong, Y., Dong, S., & Wang, Z. (2022). Development of a coupled genetic algorithm and empirical typhoon wind model and its application. *Ocean Engineering*, 248 (15), 110723.

Ha, S., Tatano, H., Mori, N., et al. (2021). Cost–benefit analysis of adaptation to storm surge due to climate change in Osaka Bay, Japan. *Climatic Change*, 169, 23. https://doi.org/10.1007/s10584-021-03282-y

Holland, G. J. (1980). An analytic model of the wind and pressure profiles in hurricanes. *Monthly Weather Review*, 108(8), 1212–1218. https://doi.org/10.1175/1520-0493(1980)108<1212:AAMOTW>2.0.CO;2

Holland, G.J., Belanger, J, I., & Fritz A. (2010). A revised model for radial profiles of hurricane winds. *Monthly Weather Review*, 138(12), 4393–4401.

Jelesnianski, C. P. (1965). A numerical computation of storm tides induced by a tropical storm impinging on a continental shelf. *Monthly Weather Review*, 93(16), 343–358.

Jung, W., Taflanidis, A. A., Kyprioti, A. P., Adeli, E., Westerink, J. J., & Tolman, H. (2023). Efficient probabilistic storm surge estimation through adaptive importance sampling across storm advisories. *Coastal Engineering*, 183, 104287.

Kennedy, A. B., Wirasaet, D., Begmohammadi, A., Sherman, T., Bolster, D., & Dietrich, J. C. (2019). Subgrid theory for storm surge modeling. *Ocean Modelling*, 144, 101491.

Kyprioti, A. P., Adeli, E., Taflanidis, A. A., Westerink, J. J., & Tolman, H. L. (2021). Probabilistic storm surge estimation for landfalling hurricanes: Advancements in computational efficiency using quasi-monte Carlo techniques. *Journal of Marine Science and Engineering*, 9, 1322. https://doi.org/10.3390/jmse9121322

Lee, Y. H., Park, S. K., & Chang, D.-E. (2006). Parameter estimation using the genetic algorithm and its impact on quantitative precipitation forecast. *Annales Geophysicae*, 24, 3185–3189. https://doi.org/10.5194/angeo-24-3185-2006

Lemieux, C. (2009). *Monte Carlo and Quasi-Monte Carlo sampling*. Springer Science & Business Media.

Marks, F. D., Kappler, G., & DeMaria, M. (2002). Development of a tropical cyclone rainfall climatology and persistence (RCLIPER) model. Preprints, *25th Conference on Hurricanes and Tropical Meteorology*, San Diego, CA, American Meteorological Society, 327–328.

MLIT (2021). Project PLATEAU by MLIT, https://www.mlit.go.jp/plateau/.

Pradhan, A., & Kim, Y. T. (2015) Application and comparison of shallow landslide susceptibility models in weathered granite soil under extreme rainfall events. *Environmental Earth Sciences*, 73(9), 5761–5771. https://doi.org/10.1007/s12665-014-3829-x

Schloemer, R.W. (1954). Analysis and synthesis of hurricane wind patterns over Lake Okeechobee, Florida. Hydrometeorological Report, No. 31, 49 pp.

Simpson, M., Padilla, L., Keller, K., & Klippel, A. (2022). Immersive storm surge flooding: Scale and risk perception in virtual reality. *Journal of Environmental Psychology*, 80, 101764.

Taylor, A., & Glahn, B. (2008). Probabilistic guidance for hurricane storm surge. *Proceeding of the 88th AMS Annual Meeting*, New Orleans, Louisiana, USA, 20–24 January 2008, Retrieved 16 May 2023, from https://ams.confex.com/ams/88Annual/webprogram/Paper132793.html, 2008.

Takagi, H., Li, S., de Leon, M., Esteban, M., Mikami, T., Matsumaru, R., Shibayama, T., & Nakamura, R. (2016). Storm surge and evacuation in urban areas during the peak of a storm. *Coastal Engineering*, 108, 1–9.

Tuleya, R. E., DeMaria, M., & Kuligowski, R. J. (2007). Evaluating of GFDL and simple statistical model rainfall forecasts for U.S. landfalling tropical storms. *Weather and Forecasting*, 22(1), 56–70. https://doi.org/10.1175/WAF972.1

You, S. H., Lee, Y. H., & Lee, W. J. (2012). Parameter estimations of a storm surge model using a genetic algorithm. *Natural Hazards*, 60, 1157–1165. https://doi.org/10.1007/s11069-011-9900-y

Zhao, M., Yue, T., Zhao, N., Yang, X., Wang, Y., & Zhang, X. (2015) Parallel algorithm of a modified surface modeling method and its application in digital elevation model construction. *Environmental Earth Sciences*. https://doi.org/10.1007/s12665-015-4177-1

Index

Note: **Bold** page numbers refer to tables and *italic* page numbers refer to figures.

For Product Safety Concerns and Information please contact our EU
representative GPSR@taylorandfrancis.com
Taylor & Francis Verlag GmbH, Kaufingerstraße 24, 80331 München, Germany

www.ingramcontent.com/pod-product-compliance
Ingram Content Group UK Ltd.
Pitfield, Milton Keynes, MK11 3LW, UK
UKHW021123180425
457613UK00005B/200